Praise for
Four-Star Leadership for Leaders

"Your efforts to help us all share the leadership principles and ideals of these distinguished gentlemen are to be commended. As you know, our ships, airplanes, tanks, and weapons all change constantly, but good leadership is a timeless commodity that no force can survive without."
—John H. Dalton
Former Secretary of the Navy

"The study of leadership is one which is crucial in both military and corporate settings and there can never be enough written on it. The insights provided by your subjects are timeless."
—General Dennis J. Reimer
Former Chief of Staff, US Army

"I think you have chosen well the testaments you present from these unique men—and they add up to extremely useful perspectives and advice."
—General Russell E. Dougherty
Former Commander in Chief,
Strategic Air Command

"It is certainly a 'tremendous' compilation of advice from the top military flag officers of our generation."
—General Charles C. Krulak
Former Commandant of the Marine Corps

"The principles of leadership set by these men and others have served us well throughout history. How well we lead depends in large measure on the way we prepare ourselves to motivate and inspire those in our charge. Learning the lessons of those who have served

before us is fundamental to our success as leaders. Your book should help capture some of those lessons."
—Admiral Jay L. Johnson
Former Chief of Naval Operations

"I am always inspired when reading about those who have contributed so much to our Nation, and believe we should strive to learn all we can from their experiences."
—General Joseph W. Ralston
Former Supreme Allied Commander
Europe

"Clearly, you have made a major contribution in presenting the qualities needed in leadership in the military."
—General David C. Jones, Ret.
Former Chairman, Joint Chiefs of Staff

"I personally believe role models are the most influential means of teaching/learning our profession. I am intrigued by the variety of individuals you interviewed, and I know their thoughts and insights will make for a thoroughly enjoyable and profitable read."
—General David A. Bramlett
Former Commanding General,
US Army Forces Command

Four-Star Leadership for Leaders

Four-Star Leadership for Leaders

Interviews with Distinguished Generals and Admirals

by
R. Manning Ancell

Foreword by
Charles "Tremendous" Jones

Four-Star Leadership for Leaders

Published by
Tremendous Life Books
206 West Allen Street
Mechanicsburg, PA 17055
(717) 766-9499
(800) 233-2665
(717) 766-6565 (Fax)
www.TremendousLifeBooks.com

Copyright © 1997 and 2011 by R. Manning Ancell

ISBN 978-1-936354-16-0

All rights reserved. No part of this book may be reproduced in any form without permission in writing from the author except by a reviewer who wishes to quote brief passages in connection with a review written for inclusion in a magazine or newspaper. Proper credits required.

Printed in the United States of America

Contents

Chapter 1
"From my perspective a good leader can't be selfish"..........15
General Walter E. Boomer, U.S. Marine Corps
April 6, 1994

Chapter 2
"You don't have any friends as CNO"......................25
Admiral Arleigh A. Burke, U.S. Navy
October 14, 1974

Chapter 3
"You're never going to prove it until you do it"..............35
General Jack J. Catton, U.S. Air Force
September 17, 1987

Chapter 4
"By Example Was My By-Word".........................41
General Raymond G. Davis, U.S. Marine Corps
December 9, 1975

Chapter 5
"I've always believed that reading is an important
part of who I am."......................................51
General Martin E. Dempsey, U.S. Army
June 16, 2010

Chapter 6
"I was more of a follow-it-up-yourself man"................63
General Alfred M. Gruenther, U.S. Army
October 17, 1974

Chapter 7
"You've got to get out and see things yourself.".............71
General Thomas T. Handy, U.S. Army
May 1, 1975

Chapter 8
"You get the right man in the job quickly." 81
Admiral Ronald J. Hays, U.S. Navy
August 22, 1986

Chapter 9
"I walked out of the Pentagon my own man." 87
General Harold K. Johnson, U.S. Army
June 4, 1976

Chapter 10
"Integrity is among the most necessary of qualities." 99
General William A. Knowlton, U.S. Army
October 29, 1986

Chapter 11
"Tell people what has to be done and not how to do it." 109
General Frederick J. Kroesen, U.S. Army
October 21, 1987

Chapter 12
"Never make a decision until you have to." 119
General Curtis E. LeMay, U.S. Air Force
August 16, 1976

Chapter 13
"Be brave-smart and not brave-dumb." .133
General William J. Livsey, U.S. Army
June 30, 1992

Chapter 14
"A good commander is also a damned good manager." 143
General William F. McKee, U.S. Air Force
October 18, 1974

Chapter 15
"Even in retirement mentors continue to play a role." 155
General Edward C. Meyer, U.S. Army
March 24, 1991

Chapter 16
"I probably would not have stayed in the Navy if it wasn't
for Admiral Zumwalt.".................................163
Admiral Paul David Miller, U.S. Navy
January 28, 1997

Chapter 17
"You'll never be a leader if you don't tell the truth."........173
Admiral Thomas H. Moorer, U.S. Navy
June 3, 1976

Chapter 18
"Persuasive and exemplary leadership is most effective."......183
General Carl E. Mundy, Jr., U.S. Marine Corps
April 7, 1994

Chapter 19
"You can be successful in peace and a failure in war.".........195
General Matthew B. Ridgway, U.S. Army
September 24, 1977

Chapter 20
"Soldiers are not intimidated by generals any more."..........205
General Roscoe Robinson, Jr., U.S. Army
October 24, 1987

Chapter 21
"Let 'em lie!"..221
General David M. Shoup, U.S. Marine Corps
May 2, 1973

Chapter 22
"Luck goes all through this business.".....................231
General Maxwell D. Taylor, U.S. Army
October 21, 1974

Chapter 23
"I was the best in the business.".........................241
General Otto P. Weyland, U.S. Air Force
February 11, 1978

Chapter 24
"I was almost killed in Vietnam." 251
General John A. Wickham, Jr., U.S. Army
September 5, 1990

Chapter 25
"You must set your standards high." 261
General Louis L. Wilson, Jr., U.S. Air Force
October 22, 1986

About the Author 265

Foreword

There is nothing nobler than a leader who has stepped to new heights of greatness through the respect of subordinates, and nothing more honorable than the act of selflessly offering up the service of one's life in order to preserve a higher good. The profession of arms has filled the annals of history with some of mankind's most powerful and poignant lessons. That alone, however, is not why people of all times and talents have continuously turned to it as the greatest laboratory where one can observe and master the art of leadership.

Much of military doctrine is founded on the premise that, although its ranks form the backbone for waging the art of war, peace is its profession. This somewhat ironic statement is the basis for one of the strongest leadership principles known to man and had been used long before the advent of military technology and standing armies. The military has always held the belief that in order to succeed, you've got to prepare for the worst. For them, it's war; for the businessman, it could be bankruptcy; for the spouse, it could be the end of a marriage. Whatever the unthinkable, the most effective basis for deterrence is found in preparation. A leader, military or otherwise, must imagine the unimaginable and take it back from there. The United States could not have fathomed total nuclear holocaust during the years of the Cold War, and yet the Cuban Missile Crisis brought the unthinkable to the brink of reality. The United States emerged as a world leader able to handle crises of global proportions through preparation. Its leaders shouldered the responsibility of, and accountability for, the future of a nation. So whether the leader is responsible for a country, a company, or a family, these same principles of leadership apply.

The second principle that makes the military such a powerful laboratory for leadership is the oath of allegiance its members swear or affirm before they enter the service. Many professions hold such an oath of service—the medical profession, the profession of law, the institutional vows of marriage. But military leaders are unique because they alone are measured by how well they motivate the tens, or hundreds, or thousands of personnel under their command. If they don't do this well, the results can be catastrophic. The military professional capable of motivating and disciplining only himself or herself can't serve as an effective leader in the military. The leader must simultaneously mentor, inspire, discipline, and love. When dealing with the preservation of national security and survival of allies, history has shown that there is no more powerfully unifying or divisive event than warfare. The military professional has got to lead, or suffer consequences of potentially global proportions. With the tremendous advent of modern society and its complicated interweaving, the military leader cannot afford to hold his or her position as the weak link in the armor.

The student of history can endlessly examine the causal events and historical accounts of the military, but for the student of leadership, the personal words and witness of proven leaders melded in the fires of the profession of arms offers a rare glimpse directly into the mind of greatness. Historical accounts, no matter how accurate, are fraught with subjective and emotional reactions, misinterpretations and mindsets. Rather than that standard historical account, this book offers numerous interviews with living and recently deceased military general officers. In essence, we get the rare opportunity to learn the secrets of history and to master the lessons of leadership from the men of greatness that made them happen.

A final word of clarification is called for here. If you've gotten this far but decided not to continue because you think this book is relevant only to people with some sort of military background, forget it. Leadership cuts across the midsection of society like nothing else and touches anyone who leads and, yes, follows others. It is my hope that no matter what your line of work or business, you can take some of these principles from the greatest leadership laboratory in the world and apply them successfully to your life.

Charles "Tremendous" Jones

General Walter E. Boomer

A decorated leader in Operation Desert Storm, Walter E. Boomer was born on September 22, 1938, in Rich Square, North Carolina. Following graduation from Duke University in 1960, where he earned a B.A. and a commission as a Marine lieutenant, he served at Camp Lejeune until assigned as a company commander in the 4th Marines in Vietnam from 1966-1967. He returned to Vietnam as an advisor in 1971. Promoted to brigadier general in 1986, Boomer led U.S. Marine Forces in the Central Command during Operation Desert Shield and Operation Desert Storm. Promoted to four stars as the Assistant Commandant of the Marine Corps in 1992, he retired the summer of 1994 and entered the private sector. General Boomer resides in Connecticut.

1

"FROM MY PERSPECTIVE, A GOOD LEADER CAN'T BE SELFISH"

A conversation with General Walter E. Boomer, USMC
April 6, 1994

General Boomer, I've been talking about leadership with people for many years. It's a fascinating but extremely complex thing. As an outsider looking in on the military service, it appears to me that there are a very finite number of superior leaders mixed in with a lot of other people who are not leaders. Why is that?

I think to some degree it may have something to do with being selfish or unselfish. From my perspective, a good leader can't be selfish. Now that doesn't mean that the unselfish person that I'm advocating as the good leader is any less of a good disciplinarian, if that's what's called for; or can't be tough when the situation demands it. But the basic underlying fundamental premise of this unselfishness has to do with giving of yourself to your people. Caring about your people. Because if they perceive that you care, they will do almost anything for you. Many of them will die for you. You can't fake it, though. And they will come to understand very quickly, it won't take them too long a period of time, to know whether or not you care about them. Now the caring manifests itself in a whole host of ways. In combat, it means sharing their hardship. In combat, it means being up front where they are. Doesn't mean being stupid, doesn't mean sacrificing yourself unnecessarily leaving them leaderless; but it means going in harm's way with them. It translates into giving them credit—letting your people take credit for the successes that occur. You shoulder the blame for the failures. There's a certain amount of love involved. Now "brains" in soldiers and sailors get very nervous when you start talking about love. But I would dare say that most of the great leaders that I have known have loved their Marines. And I think everybody knows what I mean. I don't think the selfish person understands any of that. Or if they understand it they can't bring themselves to the level that's necessary to get people to do things for you, to want to follow you.

Yours is a very persuasive argument for the notion, long debated, that leaders are born and not made, because you can't learn to be unselfish, you can't learn to love your fellow man. That's something that you either have or you don't.

Yeah. I would subscribe to the theory that you can make a pretty good leader, but that you really can't make a great leader.

While you were growing up in the Marine Corps, whom did you classify as a great leader, somebody that you looked up to, perhaps provided inspiration to you?

I was influenced very early on by my company commander. This occurred at a time when I did not know whether I was going to stay in the Marine Corps or get out. I came in with the intention of only staying for a very short period of time, so I was at a very impressionable time in my career. A captain by the name of Willard J. Woodring, a mustang—a man who had enlisted in the Marine Corps and gotten commissioned—was my company commander. He was a man who cared, a resourceful person, even-tempered. I don't recall him ever flying off the handle. He probably did, but it happened so seldom that I've forgotten about it. We understood that he was going to do the very best he could to take care of us. He knew his job. So here's a captain who had a great deal of influence over me; in fact, probably had much to do with my staying in the Marine Corps. I was influenced later on by a man who was a major, when I first met him, by the name of Gene Bench. Gene was a very dedicated guy who became sort of a mentor to me, but he was another individual who cared and who took the time to teach me and explain things to me, as well as others. Later on he was my battalion commander in Vietnam. He was a good battalion commander. The traits that I had admired early on had exhibited themselves again in combat. One of my personal heroes has been General Bob Barrow. I use him when I talk to young Marines about leadership. He was a gentleman. I never heard him cuss or raise his voice. He had a quality about him that made you look up to him. A marvelous combat record in Korea, later on in Vietnam as a regimental commander. He was here in my office last week. Still straight as an arrow. Exudes all of those things that make you want to be around him, that make you want to talk with him, that make you want to seek his advice. And he cares.

Are Marine leaders better and more effective because Marines have a greater sense of tradition and esprit de corps, perhaps even a greater sense of following and knowing how to follow orders, than their brothers and sisters in the other services?

I think all of those things that you've mentioned contribute to good leadership in the Marine Corps. I have a great deal of respect for the other services and there have been marvelous leaders in each of our services. I must say that, collectively, I think that our officers exhibit more positive leadership. They seem to be more concerned about it. They seem to be more into this whole business of being a leader and I think it starts very early. We have a very rugged and demanding officer candidate course. And, quite frankly, if you find yourself in the middle of this course and really don't want to be there, you will not make it. We then follow that with a six-month course called the basic school. No one else does this. So the first thing right off the bat is six months of school during which time we focus on leadership, primarily. We focus on just what it means to be a good officer. And it's at that point in time that we begin to talk about sacrifice: your Marines come first. We demonstrate that in a whole host of things that have become tradition to the Marine Corps over the years. For example, officers always eat last in the Marine Corps, always. If there's one officer, he's the last one to eat. If there are ten officers, those ten are the last ten to eat. So, as a result, I think we have instilled in our young officers very early this sense of what it's all about to be a leader. Then from that point on they go to their specialty school. For example, if a Marine officer was going to be a tanker, he would go through the officer candidate course, then six months of the basic school, and then he would go to his basic school to learn how to be a tank officer. So we make that investment right up front.

At what point in your career were you suddenly faced with a situation where you had to really pull together all of your leadership strengths and perhaps even to ask yourself, what do I need to do in this situation to really turn this around or really make this happen?

I think to a degree it may have been instinctive, but it was helped by good and wise counselors along the way. For example, when I was a platoon commander—when I was a lieutenant—we deployed to Guantanamo Bay. This was when tensions were running high in that part of the world. In fact, it was before the missile crisis, but during the build-up to the crisis. So our company was there at Guantanamo separate from our battalion. We were building fortifications with picks and shovels, digging trenches. We did that almost every day. My concern was that the Marines were going to get bored with that, and number two, they were going to get tired of it because it was darn hard work, so how do I keep them motivated? Well, I had a very good platoon sergeant and what we did was a very simple thing, but I think a very important thing. We took our turn at the shovel just like everybody else did in the platoon. They saw this. We weren't standing around with our hands in our pockets watching them dig. We were down with them digging. Not every minute of the day, but we were always there with them helping them build with our hands as dirty as theirs were, as hot and as miserable. That took my platoon through that period. So there have been times like that all along the way that I've been challenged. I would be the last one to say that I rose to the occasion every time. But I think in a Marine Corps career you can't go through it without being constantly challenged.

As you know, ours is a very permissive society where virtually everything is forgiven. How difficult is it to keep that very thin line between the commander and his troops without developing the relationships that then become those that could best be described as fraternization?

I think it's relatively easy. I've never perceived that to be a problem. I was taught very early on that you don't operate on a first name basis in the Marine Corps, between officer and enlisted. You don't put yourself in social settings which lead to situations that become very difficult. You avoid those. Now some might say, well, how does that tie into your thesis of caring? Very easily. Number one, that's really not what your Marines, in our case, expect of you. They don't want to go out drinking with you. Number two, they really are uncomfortable calling you by any name other than your rank. So you don't need to fall into those traps. Going back to my original premise that they perceive that you care about them, those kinds of things aren't important to them. What's important to them is that you know your job and that you're going to take care of them and if they've got to go into combat, you're going to figure out what the heck it is that has to be done so they can come back home alive. That's what they want from you. They don't want to be babied by you, they don't want to be coddled, they don't want to be on a first name basis with you. They really do thrive on this whole concept of mutual respect. They want you to respect them for who they are. If you do that, they'll respect you for who you are. So this whole thing of fraternization is not as big a problem as some people make it out to be.

Were there times during Desert Storm where you were able to observe young officers who rose to the occasion and exhibited leadership strengths only because of the situation that they were suddenly thrust into?

We expect a certain level of leadership competence. We expect that to be exhibited, demonstrated. And, yes, we had lots of young officers, lots of young staff NCOs and NCOs who just rose to the occasion, who exhibited a lot of courage, calmness under fire. So as a result, they stood out and I would like to think that we were able to reward those whom we know about. But that has always been the case. You always go into war with the majority of your officers having never experienced combat. This was true in Vietnam, with the few exceptions of those who had been in World War II or Korea, and they were beginning to fall by the wayside. So that was our first time in combat. This time, during Desert Storm, many, many of the Vietnam veterans had long-since retired. So each time you take a whole new batch of really untested Marines into battle. Then the test is whether or not they perform.

I remember General Shoup once said that he believed that great leaders come about because of the willingness of the people they are leading to be led. Is there some truth to that?

I think there's some truth to that. To a degree a bit of that is instilled at Parris Island at recruit training and at Quantico at officer candidate training. This whole thing of followership is emphasized during those periods. I think for the most part it sticks with Marines throughout their career, because during those two periods of time in particular, unless you're willing to follow, unless you're willing to shed a bit of yourself for the common good, you won't make it through.

I think I picked up, General, early in our conversation, that you felt that you were mentored. Have you been a mentor?

I would say that I became a mentor to several people about the time that I was a lieutenant colonel. By that point in time I had considerable experience, two tours in combat, battalion commander. So I began to feel that there were those that I could help, whose personality and whose leadership traits I admired. So I have tried to do that with a number of people over the years.

As warfare becomes more technological, does that infringe in any way on leadership or replace it or substitute it in any way?

No, I think it makes good leadership even more important. Just one example, and that is the whole issue of fratricide. If you don't do things almost perfectly today, the chances of someone getting hurt from friendly fire are so much greater than they were when I was in combat in Vietnam. I recall one day in Vietnam my company was moving across a hilltop, we were separated from everybody else by design, and all of a sudden from nowhere an artillery barrage came down right on top of us, right square on top of the company. It was as if somebody had been plotting our movement for hours and then very carefully had fired artillery at us. I didn't think that it could be the enemy because I didn't think they had that kind of artillery anywhere in the area. As it turned out we discovered very quickly that it was friendly fire. It was a South Vietnamese battery that didn't know that we were there. In other words, the word hadn't come down from the U.S. side, down through the Vietnamese channels, down to this artillery battery. We finally got it turned off. But interestingly enough, I only had one man hurt. Now there were a few miracles involved there, I suppose, but my point is, here is an artillery barrage right square on top of this company. I mean, right square on top of us. I only had one guy badly wounded. He was my XO. That's the way artillery is. That's the way older weapons are. Fragmentation weapons, you can come out of a tough situation unscathed. With the missile, it's a different story because the missile is going to hit. If you point that missile at another tank or another armored personnel carrier and pull the trigger, so to speak, everybody in that vehicle is going to die. And if you made a mistake, you've made a horrible mistake. And we made a couple of those mistakes during Desert Storm. So technology makes knowing your job even more important. And, of course, knowing your job is one of the critical pieces of good leadership.

Some have said there are no longer any Pattons or Chesty Pullers because senior generals are required to do so much now. They have to not only be effective leaders but they have to be part manager, part administrator. There are several other parts that make it up. Do you find that to be true in the Marine Corps today, the premise that there is no longer room for what has traditionally been known as the warrior?

I don't think that's true at all. I think you can leave this headquarters or any other office and go out and be a very fine warrior. For most of a military person's career he's not involved in combat. So as a result, today, as well as yesterday, you are required to be a manager, in some cases an entrepreneur, a whole host of things. But that doesn't, that shouldn't, in my view, affect what's in your heart and that is a warrior spirit. Now that manifests itself in a lot of ways. Some people are quiet warriors, others are more flamboyant. Either kind can get the job done. Although I must say that I've seen more quietly determined successful people in combat, than I've seen the flamboyant types.

Did you witness any exceptional leadership by the enemy in battle during Desert Storm?

No. None.

What does that tell you, General?

It tells me that they don't understand the investment that has to be made in order to be successful in this profession of arms. They're not willing to work at it hard enough. This is a very, very tough business. It requires hard work every day, whether you're in the office or in the field. If you're in the field, it requires training, day after day, month after month, year after year, and an investment in training. In my view, they don't understand how great an investment that is. Even if they understood it, they're probably not willing to make it. I was glad that I didn't see it. It made our job a lot easier. They did a few things well. They put up a barrier across Kuwait in perhaps record time. They did a couple of fairly decent engineering jobs in conjunction with that, but then when it came down to really being ready to fight us, they were not. Some of that had to do with arrogance on their part. That had to do with their culture. They had to convince themselves that they were better. So as a result I suppose that they felt they didn't need to do very much in order to prepare. Well, they were mistaken. As we watched the situation from Saudi Arabia from an intelligence perspective, it became obvious to me that over time they weren't doing any training. They never moved. Tanks were dug in and that's where they stayed for weeks. We, on the other hand, trained every day. In fact, we trained so hard that...several weeks before the attack into Kuwait I told my commanders to pull back a little bit. I was afraid that they were going to wear our troops out training so hard. But that's the difference in mentality. So we were primed and ready to go. Speaking of that barrier, we had taken that barrier, or we had taken a picture of the barrier and used that to construct a mock-up of it in Saudi Arabia. So we trained against that, day after day, because that was a concern to us to get through that barrier quickly.

One final question. On your last day in uniform, what words are you going to leave with your successor and your fellow generals and fellow Marines on the subject of leadership?

I've thought about that. I think what I may do, and I haven't told anybody this, is I may have at that final farewell a brand new Marine from Parris Island there. Maybe I'll have a male and a female. And what I will say to them is, if in your decision-making process you keep these two youngsters foremost in your mind, you won't make a wrong decision. If your decisions are based on what's good and necessary for them, they will be good decisions. Not only at the headquarters level, but at the platoon level.

Admiral Arleigh A. Burke

"Thirty-One Knot" Burke, the nickname he earned for his exploits as a daring destroyer captain in World War II, was born on October 19, 1901, in Boulder, Colorado. A graduate of the Naval Academy in 1923, he alternated assignments with the Bureau of Ordinance and at sea. Following World War II he was chief of staff to Admiral Marc Mitscher, commander of the Atlantic Fleet, earning promotion to rear admiral in July 1950 despite the controversy of his involvement in the "revolt of the admirals." In August 1955 Burke became Chief of Naval Operations, propelled by President Eisenhower to four stars over the heads of ninety senior admirals. He served an unprecedented six years as CNO, retiring in August 1961. In 1983 a new class of destroyers was named in his honor. Admiral Burke died on New Year's Day, 1996.

2

"YOU DON'T HAVE ANY FRIENDS AS CNO"

A conversation with Admiral Arleigh A. Burke, USN
October 14, 1974

Admiral, toward the end of the war you were on the staff of Vice Admiral Mitscher, an aviator who didn't care much for surface ship drivers. How did you fit in?

When I went there he did not want me on his staff at all. He had a good chief of staff, an aviator, and he wanted to keep him. I came aboard at sea when I came up with my destroyer squadron someplace north of the Solomons while he was on the way to make the first carrier strike west of Truk. I came aboard the Lexington by highline, went right up to the flag bridge, and reported who I was. He said, "Where's your gear? Is your gear aboard?" I said, "No, sir. I don't have any." All I had was on my back and my shaving gear. He said, "You've got to go down and take a good hot bath. Go down to my cabin and turn in. I know you're tired—it's been a long time since you slept in a bed." And it was, too. So I went down and took a shower and sent for some clothes—khaki uniforms and under clothes. I thought, well, damn—I'd really like to sleep but I felt if I turn in I've had it. I dressed and went back up to the bridge and reported in. Of course you don't ask him what you're supposed to do—you're supposed to know what to do. He just grunted, so I talked with other members of the staff to find out what they were doing. Truman Hedding, the man I was relieving, was not aboard. I went over to give Admiral Mitscher an opportunity to see me if he wanted to or call me over, but he didn't. He wasn't going to talk to me—he didn't say anything. This went on for about three days. I read everything I could think of that might be useful. I talked with aviators. After I worked on that he gradually came to trust my judgment and depend on me, but he never did like non-aviators. He tried to get me to go to Pensacola right after the war. "Admiral, I don't want wings. I'd be an ersatz aviator. I wouldn't be any good."

He sounds like a tough fellow to work for.

He was very understanding of his men. He drove his men hard. They performed or else. I was his hatchet man and when a flag officer or senior officer did not perform it was my duty to get rid of him and go tell him, "Sorry, you've had it. You have to go back to the states."

Where does a flag officer who's just been canned go?

Out. He's had it. War is that way. It may be beyond his capability, but that's too bad. When you permit improper performance of duty you accept an excuse and excuses get poorer and poorer and people don't try very hard. This is why it's absolutely necessary that you have very high standards and you make sure those standards are met or you cut throats.

What did you determine to be your number one priority as you took over as CNO?

I realized that my most difficult job was to obtain the maximum effort out of the senior officers toward a common direction. Sometimes you lose a friend, but you usually get him back. You have to make choices. One of my friends is an excellent admiral—four-star admiral. He felt he should have had a job that I gave to someone else. The other man did a very good job but you have to make a choice. I did what I thought was right but I could be wrong.

Lots of times there are things that other people know and you don't know and you make the wrong choice. But you've got to make it based on what you know and not what somebody else knows. You've got to make sure you know as much as you can but neither can you wait forever until you find out. You lose sometimes.

Did you receive the cooperation you needed from the senior officers in order to make things happen when they needed to happen in the Navy?

One of the early things I learned was that I could not get anything accomplished by giving an order because I didn't have the time to follow through on that order. What I could do, though, was convince somebody that this was a good thing to do and get him to follow through on it.

Wasn't there resentment among the admirals who were senior to you before you became CNO?

Sure, they were all resentful. Why not? No man reaches a position like that without feeling he is qualified. They're all good people. They all spent a lifetime—thirty, thirty-five years working. Of course they resented that they had been passed over. Now there was my biggest job—to get rid of that resentment. You can't order it out. It's natural. It's there. They can pat you on the back and be nice but still resent it on the inside. They aren't going to perform until they don't resent it inside. Your destiny is dependent on those admirals. The most important thing I had to do was get their confidence that what was being done were the correct things to be done for the Navy—for the country. They had to believe that. No organization can go very far unless it has a cause that's bigger than itself.

In the early weeks of your watch how did you choose new staff people?

I didn't choose any staff.

You kept all of Admiral Carney's staff?

Yeah, complete—except for one man, his flag lieutenant. He was due to go. This could have been a mistake—I don't think it was—but this one young fellow was due to go so I looked down the list. I had a list of all my destroyer captains. The flag lieutenant is very close to the admirals. He has to know how he thinks, what he's going to do. He has to act for him lots of times, so it ought to be somebody with the same sort of background. I chose my flag lieutenant from the list of captains of COMDESLANT, that I'd just been with, and I chose a man I had never met because he had such a hell of a good reputation. I had one ship per page and every time I heard about the ship I put down what I heard. I'd pick up bluejackets. "What ship are you from? How good is it?" Sometimes they didn't know who I was, sometimes they did. Everything I could get on the ship I would have. So I looked down there and picked Weschler. He turned out to be a hell of a man. He's commander of destroyers now in the job I used to have. But he was the only one. All the other staff stayed. The staff were picked by Admiral Carney. They were loyal to him. They were used to doing it his way. They thought he had gotten a raw deal, a dirty deal and I was a usurper—knowing damn well it wasn't my fault, I didn't do it, but still I was a usurper. They resented the hell out of me but they were experienced and they were loyal to the Navy. I called them in and said, "Now, look, I don't give a damn whether you like me or not. That's not the point. What I'd like you to do is the very best job you can. Advise me." That group of people never did like me like they liked Admiral Carney—you couldn't expect it. But they did a good job. I had a group that was experienced but a group I had to always be careful with. They didn't give a damn if they were pleasing me or not. They were operating strictly from a duty proposition and that's good. Uncomfortable, but good.

Admiral, taking on a new, larger assignment with increased responsibilities really puts someone's leadership skills on the line. When you moved up in rank and responsibility did you employ a particular strategy?

You can't have a personal strategy. You can contribute, more or less, to your total effort, but when a man tries to build himself, or to impress his own views upon the service, or his unit, or his organization—whether it's in the service or not—usually he's wrong. It isn't that his ideas are altogether wrong—they're usually pretty much right—it's that he didn't know how to do it. The first thing he's got to have is a cause. It's got to be bigger than he is. Now, within that is what he's best capable of doing. He's got to have a cause bigger than himself. Nobody was attracted to John Paul Jones because of his being a damn good sailor. They were attracted to him for what they believed in, what they wanted to see accomplished. He was skillful enough—a professional—and they knew that.

Can you quantify the factors that made you a capable and effective leader throughout your naval career?

Being immodest about it, I was a professional. I knew my trade. I knew the limitations, as well as the capabilities of the equipment and the people. I knew that you can get a lot out of men if you inspire them to do their very best. First you have to tackle what is reasonably possible. Timing has a great deal to do with it. I was fortunate that I felt the time for missiles had come to the Navy. I felt the time for nuclear power had come for the Navy. And we were able to develop these things. The people in the Navy recognized that first—this was for the good of their profession and organization and they were permitted to contribute their maximum to it and they did. A good deal of this is timeliness. Also a great deal of it is knowing how to get people to do their very best.

Are you telling me, Admiral, that throughout your career the timing was just right for all the events that occurred and you just happened to be there?

Yes, a good deal of that. I pushed it, maybe, a little faster than someone else would have done. Really, I just happened to be there. Now this is true in battle, too. I made a pretty good reputation in combat and a surprisingly good deal of that was being there at the right time and the right place. The thing that I had was the ability to use that time and that place. I had the skill to do that, but someone else could have come along and done just as well—maybe even better. It's not just happening to be lucky but luck does play a part in it.

It's my understanding that you had no tolerance for mediocrity.

That's true.

How did you handle the cases of people who weren't pulling their weight?

I got rid of them.

Summarily?

Sure. The first thing that a commander must learn is not to tolerate incompetence. As soon as you tolerate incompetence—it doesn't matter why—you have an incompetent organization. It's quite natural that the level of performance of an organization always goes down if you have mediocre people. It's quite natural. It doesn't take very many mediocre people—if you permit it. I'd rather have that incompetent person near me if I'm going to tolerate incompetence at all. I'll put the competent person way out where he makes decisions and I don't have to watch over him. It's not good to surround yourself with competent people—that doesn't do it. You must put competent people in your organization in the right spots. You'll make mistakes. Some people can't do it. Some people are not equally capable of all things. You can't assume that a man is suitable for all types of work, so you make mistakes.

The mark of success of a good businessman, I learned after I retired, is knowing when to cut his losses. In the service you've got projects that cost tremendous amounts of money. You have to keep checking on those all the time to see whether they will be successful. If it's not, for God's sake don't waste any more money on it. The same thing is true of people. You have to be very careful with people, to develop their potential. You aren't going to get a man who has been a follower all his life to suddenly be able to make correct decisions without having made decisions before. So you have to train people to make correct decisions, which means making decisions and living with them. You don't know if they will be successful; sometimes they're not. Sometimes something happens or you make a misjudgment. This is true in projects. Polaris was tremendously successful, but the SeaMaster—which started out successful and looked like a very good idea, a beautiful seaplane—was a flop. The concept was good, but technically it was a flop. So I tried to kill it. There was no reason why this thing couldn't do what it was supposed to do. It was people; I thought they didn't have the right people, and it turned out they didn't. They weren't good enough; it was mediocre people they put on the project. One extraordinarily good seaplane man could have pulled it out.

When Polaris came along, I wanted to make damn sure I didn't have any mediocre people there. That's why it was successful, not because it was a beautiful concept, but because they had the people in it who were technically qualified and able to perform.

How did you reward people who consistently overachieved?

Give them credit for what they did. Nothing more. It's hard to do, because a lot of people have a lot of bosses who say, "Look what I did." People like that have people who don't like to perform for them—too much of the personal pronoun. They didn't get credit when they deserved it. Mostly a man wants recognition—by his peers, by his family. There can't be any baloney about this, either; it has to be real.

Admiral Burke, one of the great challenges of leadership is picking good people and putting them in the right positions. How did you accomplish this?

Think among your associates—people you know. You can eliminate 75% of them for any position you know a little bit about—probably more than that. Your field comes down pretty fast. You may feel sorry for 'poor old Joe,' and let him have a crack at it, but if you do the chances of his failure are much greater. That's one thing you've got to avoid like poison. I had to tell my very best friend he was finished—

he couldn't hack it. That didn't mean I didn't like him personally—I did—but he didn't have it. I said, "Sorry, Joe," and he said, "Dammit, you're not giving me a fair shake." Maybe you don't. Maybe you have to lean over a bit too far backwards, but you've got to do that so that the good people will realize it all has to do with excellence, not who you know. It's not trial and error; a lot of people have gone before. You can take all the great leaders—at least in the military—and study their lives to see why they were successful: zeal, a willingness to work, mental capacity. Pretty soon you can pick people pretty well.

I worked my people very hard, particularly my immediate staff. They worked like hell. That's another thing. You've got to work as hard as your people. Mostly you have to know what you're doing, and be a professional. Professionalism throughout the Navy is what I strived most for. A naval officer—a line officer—should be a professional in combat at sea. He may have a lot of other qualities that might be helpful, but basically he's got to be a good combat officer. That doesn't mean that everybody—every line officer—must be fired if he's not a good combat officer. It means that if he isn't a good combat officer he's got to have some damn superior qualities in other ways to make up for his weakness.

You don't try to treat people equally; there's no way you can. People should be treated for the good of the organization. You've got just one thing to worry about and that's how well the job can be done. You can't worry about yourself, about your time at home, about your wife. You've got one job and it's the most important job you'll have in your whole damn life. Sure, you'll lose some friends, but a man who worries about his friends in a job cannot do a good job. You don't have any friends as CNO. You can't. You hurt people; you've got to hurt people. You've got to give preference to those who can perform and do. You've got to give credit to the people who try to perform, and can't, but you can't put them in positions of responsibility. The worst people of all are those who could, and don't. Those are the people who have the capability but will fold on you in a moment of crisis. You must get the very best people you can find who can perform.

General Jack J. Catton

Born in Berkeley, California, on February 5, 1920, Jack Joseph Catton dropped out of Loyola University in Los Angeles to join the Army Air Forces in 1940, earning his wings later that year. Following duty in the Pacific during World War II he served in bomb squadrons and wings and saw duty on several occasions at SAC headquarters. In 1959 Catton was promoted to brigadier general, the youngest general or flag officer in the Armed Forces. He earned a fourth star in 1969 as commander of Military Airlift Command and headed Air Force Logistics Command from 1972 until his retirement in 1974. Thereafter he was active in the aerospace industry until his death on December 5, 1990.

3

"YOU'RE NEVER GOING TO PROVE IT UNTIL YOU DO IT"

A Conversation with General Jack J. Catton, USAF
September 17, 1987

Is it fair to say that for the most part when you look at the Air Force over the last forty years, and perhaps the Army over the last hundred-some years, that in terms of its senior leadership it doesn't change a whole lot, that the people element is virtually the same now as it might have been a hundred years ago?

I think that good leaders, successful military leaders, are cut from the same piece of cloth now as they were before. The elements of leadership or the characteristics a person must have to be a successful military leader aren't going to change very much.

Can the military service itself take credit for that?

Yes, I think that military services can take credit for that, but I think that the facts or the demands on a military leader are so much the same in spite of advances in technology. I would guess that General Lee had much the same characteristics as General Eisenhower, much the same leadership characteristics as General LeMay. I think the elements of leadership, the characteristics necessary to be a successful military leader, are fundamentally the same, although personalities can be terribly different.

General, would you agree that today, in 1987, somewhere in the Air Force there are the LeMays and the Powers and the Cattons just waiting for the right opportunity?

I would certainly expect that to be the case. You know, for example, we go back and talk about LeMays and Pattons and Eisenhowers and Bradleys and people like that in World War II, and I guess what you can say is the military leaders have not emerged as nationally recognized military leaders and heroes because we haven't won a war since 1945.

That's one of the great difficulties the military has today. We don't have any recognized military spokesmen. When I say recognized, I'm talking about recognized by the people of the United States as an expert, a proven military expert. For example, in the Air Force, name me one after LeMay.

That's pretty hard.

You can't do it. Name me one in the Army. Name me an Army general that has made the case that he is the expert, that he is the senior military statesman, that he should be listened to by the people of the United States and the congress of the United States. I don't know of any.

You have to have some combat of consequence, you have to have opportunity for these people to prove themselves and emerge. And you're not going to do that short of a significant conflict and you've got to win. Nobody emerged out of Korea. Certainly nobody emerged out of Vietnam except for Bill Westmoreland who got more hammers than he got help. So, anyway, I think, yes, the answer to your question is that I am confident, then, the military leadership that this country would need is there. You're never going to prove it until you do it.

General Catton, how would you describe your particular leadership style? Was it any different from your contemporaries?

Well, I like to think so. When I was a three-star, for example, and commanding SAC's 15th Air Force at the time, I was asked to give a lecture on leadership at our Air Command and Staff School down at Maxwell. That took so well that I was asked back many times to both the Squadron Officer's School and the Command and Staff School and the Air War College. In other words, the military education effort down at Maxwell at our Air University actually used my comments, which of course were taped, as text. And they still use it as a text down there on leadership. Now then, was my style of leadership different from others? I think that it certainly was a lot alike. There were lots of likenesses in my style of leadership, perhaps like General LeMay, perhaps like General Sweeney, perhaps like others who were partial role models for me—people that I watched and worked for. You have a tendency to pick the things you think they're doing best and try to emulate them. I would say the composite becomes a style of leadership that's a little bit different than others. But in my case, I think I was fortunate because I was young enough to be a very active participant in not only the planning and training, but also in the execution of missions. I was able to do that and as a consequence of that I really had a very thorough understanding of what my resources could do in terms of performance. I'm talking about weapons systems, I'm talking about my people. I'm talking about my support systems. I really believe—and this sounds awfully cocky now—but I really believe that I had a very good grasp of what could be done with what I had as resources—limitations as well as what we could do and how we could excel, in other words.

And, I suspect, certainly not unwilling to employ them to the fullest degree when you thought it was absolutely necessary?

Oh, absolutely. So the style of leadership, I think, was well accepted by my people because I communicated with them on a personal basis just as often as I possibly could. I participated in all phases of the command's responsibilities as much as I possibly could so I could have a good understanding of what they were facing and what tools they had to meet the job or get the job done. Compassionate and, at the same time, demanding discipline and mission-mindedness. I think I got to my people pretty well through a very definite effort at personal contact, even if it were through audio-visuals—when we got audiovisual capability. But I was on the platform talking to lay people just nearly every day. I think that helped.

You mentioned General Sweeney. It occurs to me that there were some parallels and probably some means of comparison between the two of you. Am I too far off base to suggest that perhaps he was just a little older version of you, that you shared many characteristics?

Gee whiz. I wouldn't put myself in that class. I think that we were a lot different, although he was a major role model for me. He had a brilliant intellect. He was absolutely fair. He was very demanding. He really understood the importance of his people in terms of selecting the right people to do the important jobs. He really spent a lot of his time on personnel matters and I learned that from him, surely. Going about your business in a way that people could admire—in other words, not being a loudmouth, not being overbearing, but being a disciplined gentleman about the way you do your military business. I learned from him.

How do you remember General Sweeney?

I would say that he was a handsome, physically-fit gentleman—reserved. He certainly would have been the chief and perhaps the chairman if John Kennedy had not been killed. That's what I think. And, of course, Cam Sweeney died of pancreatic cancer way too early. But Cam Sweeney would have made a wonderful chairman, whereas General LeMay, for example, would not make a good JCS chairman. General LeMay needs to be leading troops. Really, his best suit is as a military leader in command of a fighting organization. His abilities in the political areas—in knowing how to deal with matters that need negotiation and compromise—he did not have those cards. Sweeney did.

General Catton, what do you think of the senior leadership in the Air Force today? The people that you'd seen and observed probably when they were colonels.

The senior leadership of the Air Force now are people that I really don't know, didn't know when I was active. In other words, I've been retired now for nearly 14 years and the people who are now the leaders of the Air Force were unknown to me when I was on active duty, nearly all of them. What I know of them now, though, people like Larry Welch and Jack Chain, I'm very impressed with them.

So once again the Air Force has managed, through whatever system that might be in place, seen or unseen, to move these people up.

Well, we have some institutional and corporate knowledge that persists. I believe that. I'll tell you really, when I retired, I was very confident that there were so many wonderful leaders in the next strata below that we were in really good shape. Sometimes that worked out, sometimes it didn't. But we have, certainly in the Air Force, we have the potential of great leadership.

General Raymond G. Davis

A recipient of the Medal of Honor during the Korean War, Raymond Gilbert Davis was born in Fitzgerald, Georgia, on January 13, 1915. Following graduation with honors from Georgia Institute of Technology in 1938, he was commissioned in the Marine Corps. He saw extensive action in the Pacific during World War II and later, in the Korean War, commanded a battalion of the 7th Marines. Davis served in Vietnam from 1968 to 1970, first as deputy commanding general of the Provisional Corps, then commanding general of the 3rd Marine Division. His final assignment was Assistant Commandant of the Marine Corps, for which he earned a fourth star. General Davis died on September 3, 2003, in McDonough, Georgia.

4

"BY EXAMPLE WAS MY BY-WORD"

A conversation with General Raymond G. Davis, USMC
December 9, 1975

As you were moving along your career in the Marine Corps did you ever have expectations of one day becoming a four-star general?

It's really not the kind of thing you sit down and plan on. I recall as an ROTC cadet in Georgia Tech many, many years ago that one of the things that a classmate said about me entering the Marine Corps was—that the highest rank the Marines have—this is back in '38—is major general, so you can never make four stars. My response was, well, I hadn't even considered that. I think that attitude prevailed throughout. Do whatever job comes your way, do the best you can, and you don't necessarily have to point toward any such goal as how many stars you're going to make.

Does that mean you just do the best you can and if it comes along, fine?

That's right. As you serve and as you see things transpire you realize more and more that this is a "that's the way the ball bounces" situation. You know from years ahead that the selection from colonel to brigadier general is something like six out of a hundred. So how can you plan on a thing like that? You can only do the best you can and realize that it's competition. Doing the best you can in my case normally meant taking care of the best interests of the troops and the service. Probably in that order.

Do you think that your ideas about that are unique?

No, I think there are extremes in both ways and a lot of people in between. I don't think anyone will say that they I have no ambition to go to the top because that means they have no ambition to do a good job. If you have no interest in what you're doing, you obviously will not go. To me and to, I think, the majority of people, it's just one of those things of accepting the challenge at the time. I don't know many who entered the Marine Corps with a driving desire to be a general officer or a commandant as a prime motivation. I just don't know many like that.

Can I assume, then, that the Corps, as it takes individuals throughout their careers, the natural course of things weeds out people of less talent and naturally propels those who have the tickets?

In the main, I think it's one of the most competitive and most effective weeding-out processes that you can imagine. From the very early days and each promotion after—when you make captain, for example, for each promotion up the ladder there's a skimming off. Initially, the lower 10 or 15 percent are skimmed off and passed over, so to speak, and removed from the service. Then, this percentage increases—up until lieutenant colonel to colonel it must be something more like 25 or 30 percent are skimmed off and retired from the service. These laws change all the time. The law I came up through was one of about six or eight percent of the colonels didn't make brigadier. The others could serve as much as 30 years and then are forced into retirement. Everybody realized from the beginning that this was a competitive up-or-out kind of situation.

What do you feel made you successful, other than the obvious: work hard and do the best that you can.

I think my career pattern was largely by accident a very favorable one in that every time a war started I seemed almost overdue to join the forces and head for the war. I was always among the first to go. I went to World War II. I went right out to Guadalcanal. Korea, I went right out to the Inchon landing and was in the Chosin Reservoir thing. These things are not things you plan, they're things that happen to you in the normal planning of rotation of duty and so forth, much of it beyond your control, except you do have to have a desire to be involved to be in these things and you have to make your desires known. When there is an opportunity, you have to tell your boss that this is exactly what you want to do and prevail in an argument with him to let you go.

So in many ways, for individuals such as yourself, being in the right place at the right time combined with having the initiative, the drive, and the ability, is almost necessary to get ahead.

Yeah. To really move out you have to have that. Another great fortune I felt I had was that on two or three key occasions I was involved with an outstanding leader with some characteristics that I could see and realize how they could promote success. Started out early with Louis Puller in his regiment in World War II, then I went to Korea with Litzenberg, who was a different kind but also one of our greats. I was exposed to Lou Walt and a number of others. Through the years I seemed not only to be at the right time and the right place, but also to be involved with some of the key history makers in their roles.

Which individual would you say most affected your career in a positive manner?

It's hard to put my finger on one. Certainly Chesty Puller, Litzenberg, Nickerson and Walt were four outstanding ones. Then in Vietnam I had exposure to some outstanding Army people. Stilwell out in Korea. Bill Rosson, who was Army commander in the Pacific. Abrams of course, down in Saigon, spent an awful lot of time up in my area and gave me a lot of guidance and also a lot of examples of the kind of leadership it takes to get things done.

Do you feel, General, that the Marine Corps tends to discourage or encourage individuality?

I guess some of both. To me a Marine is a highly individual personality who is able to subdue or to mold himself into a team effort. In other words, it's a team effort of a lot of highly individualistic efforts. That to me has been the secret of the Marine Corps. I don't think anybody ever, in my career, tried to put people into a precise shape or form. Sure you have rules because these rules are what hold the unit together. But I think the key to the Marine Corps has been taking these individuals who have great spirit and great desire and determination on their own and convincing them that the best way to get what they want accomplished is through a unity of effort, a combination of themselves into a team.

Individuality, then, doesn't really have much of a place except maybe in the individual leadership of commanders?

From commanders right down to the fire team and squad. I found great individual leadership among my four-man reconnaissance teams in Vietnam who would go out and stay in the bush, way out by themselves. Therefore the individuals who were put out there with a general statement of their mission—what they had to get done in two or three days—they got it done. I could have written a 20-page SOP and tell them exactly how to do this and there were plenty of instructions published, but each team got the thing done in almost entirely different ways. Again, the objective, the goal of the team was important. The specifics of how they did it on an individual basis was, I think, a lot of flexibility and a lot of permissiveness on the part of marine commanders.

General Davis, how did you find out, throughout your career as an officer in the Marines, what was going on in the lower ranks? How the people felt? What the pulse of the enlisted people was?

I guess it started in my early years even in college when I became a student of the history of Stonewall Jackson. The thing that impressed me most about him was that, when there was a problem, he would be on his horse going to the scene of the trouble. That pretty much describes my approach. Whether the trouble was something wrong in the barracks or something like a fire fight on a hill, I just had a drive, an urge to go there and confront and face whatever that was. This has led me to move right into the middle of every kind of situation imaginable and confront the people and talk to the people. To get their views I had a system of symposiums with junior officers that I established in various places ahead of the trend. Visiting troops in their barracks, never having an inspection without stopping and finding out where individuals were from, how they spend their liberty, what they're doing toward off-duty study, marriage, children, families, financial problems—this kind of thing. I suppose the answer to your question is direct contact with as many individuals as my time permitted and this was day, night, midnight, before reveille, or whenever there was an opportunity to come face to face and involve myself with individuals.

What was the reaction from your staff, junior officers, the enlisted people, the sergeants? Did they appreciate this? How did they generally react?

Well, you run into two or three things. Those who are having problems somewhat resent the commander's direct insertion into their problem areas. On the other hand, it gave the junior officers a chance to display their wares, for example. These were times when some mighty good things were unearthed and they were rewarded for it. I never tried to use this to try to undercut the commanders. This happens in a lot of situations where a commander gets so involved in the details of management—and has so many management reports and management spies, more or less, direct reportings and committees and whatever all through the ranks—that he disrupts the chain of command. Mine was never of that sort. It was more like the inspector general where the inspector general sits and listens directly to those in the lowest echelon and their views and their complaints, primarily. Again, this idea of a direct approach and going to the scene of the action, more or less, seemed to succeed.

General, you spent a lot of time, many years of your life, in

combat. **How did you relax under what must have been very trying circumstances?**

I don't know. I think that when the circumstances were trying, you were so busy, so involved, that you recognized the necessity to get things done and there wasn't a whole lot of relaxation. Now during the routine of combat in Vietnam and so forth, there were days you could anticipate when you'd have a little slack time. I like to hunt. I like to swim. I even went to a lot of movies in combat when there was the opportunity. Just something to unwind. Of course reading a lot and writing some. Just the things that people do to pass time when there's an opportunity.

Were you known pretty much as a strict disciplinarian?

I don't think so. I was never a shouter. A quiet positive approach. I ran into one problem in one of my wars where within three months' time I had to relieve five different battalion commanders. Now this is extremely difficult to do and not many commanders ever had to do that. I'm not proud of it, I was somewhat ashamed of it because it meant that my communications and my leadership weren't being brought to bear properly. I just don't know how you could translate that into a reputation. I don't think that anybody ever felt that they could get away with anything or run over me. But, on the other hand, they didn't expect me to throw things at them or shout at them or grab them around the neck or anything. I guess the Marines tend to foster the idea of unswerving devotion to duty and a terrific amount of discipline. We had generals called "the screaming midget," we had one called "the blow torch" and this kind of thing. I don't think I was ever considered to be in that category.

To what degree do you believe in taking risks in difficult circumstances, not only in combat but in making difficult decisions?

Well, risk-taking is subject to a lot of definitions. As far as boldness is concerned and decisive action, there was never any question in my mind. If it was time to commit all of your forces with a minimum reserve in order to win a decisive battle, there was no hesitation. I guess the same thing applies a lot in peacetime. I was involved a great deal in the education and development effort in the Marine Corps. The bold ideas just had great appeal to me. And boldness, of course, I guess by definition, involves some risk. Very many times there's not too much difference between receiving the medal and getting court martialed. If you succeed you get a medal, if you lose you get court martialed.

What qualities did you look for in selecting a subordinate either for a staff job or for a promotion away from your staff?

I think competence. If a guy had competence that was obvious, he always was upper among those being chosen. You can gauge competence from his education, training, his ability to do, to speak, to write, his physical conditioning and so on and so forth. It all adds up to a competence that you can gauge in your own mind just from observing his immediate performance and his past record.

You were assistant commandant during the Zumwalt time?

Yeah.

What effect did this period of tremendous upheaval and change have on the Marines, on the people that you saw, and how, particularly, did the Commandant react to this? I imagine you probably talked to him at length about it.

I think at both ends of the rank structure, the Commandant and I had many sessions. The whole approach in the Marine Corps at the top was if all those people are going in that direction, we'll head in the other direction and invite people who want to go our way to join us. This was successful. We found that among the American youth there were great numbers of people that wanted to go the other way and they saw us as the leaders in that direction. And this was deliberate on our part, to tighten up, to restate our standards and enforce them. Now, I traveled around some of that time and it was most interesting to go aboard a Navy ship or a Navy base where there was a Marine detachment. I found without exception that every Marine, every Marine in every detachment, seized this as an opportunity to get his hair cut shorter, make himself cleaner, stand taller, be a better Marine so that he was a man apart from what he saw going on in the Navy—almost universal among Marines serving with Navy units. To them it was just a challenge to not be a part of what they saw and to lean over the other way entirely.

General, the Marine Corps for the most part looks and behaves and performs no differently today than when it was founded more than 200 years ago, while the other services have gone through a long evolution of changes. Do you think it's time for the Corps to better reflect American society as a whole?

A short answer: no, I don't. As I see it there's only room for a certain number of elite forces in any group of forces. The elite force has to be the one that's different, the one that's a challenge to its members. As soon as you reduce these challenges and the requirements that the fellow has to meet and accomplish in order to be a member, as soon as those are diminished then you've diminished the quality of the fellow who wants to compete. So I don't think you can have a Marine Corps which, by definition, is an elite force without having these kinds of standards that require a different kind of man or woman to step forward and say I want to try to do that.

General Martin E. Dempsey

The U.S. Army's 37th Chief of Staff as of April 11, 2011, General Dempsey was born in New Jersey in 1952. He graduated from West Point in 1974 and was commissioned in the Cavalry. Most of the first 20 years of his career Dempsey held command and staff positions with troops, interrupted briefly with graduate study at Duke University, completion of the Army and General Staff College at Fort Leavenworth, and three years as assistant professor at West Point. Promoted to brigadier general in 2001, he rapidly rose to four-star general over the succeeding seven years with duty as commanding general of 1st Armored Division and deputy commander of US Central Command. He assumed command of Training and Doctrine Command (TRADOC) in December 2008 and served until his selection as Chief of Staff.

5

"I'VE ALWAYS BELIEVED THAT READING IS AN IMPORTANT PART OF WHO I AM."

A Conversation with General Martin E. Dempsey, USA
June 16, 2010

You were classmates with General David Petraeus. What do you remember of him in cadet days at West Point?

Well, there are four of us from the class of '74 who are four-star generals. It's Dave and I, and it's Keith Alexander, the general that was just appointed to command the Army's Cyber Command, and there's Skip Sharp, who is the commander of U.S. Forces Korea. Oddly enough I knew them very well. It just turns out that three of us, Alexander, Sharp and I, were in the same cadet regiment—roughly, you know, 250 cadets. Petreaus was in the first regiment, so I didn't have as much contact with him, but because we both grew up in the same county of New York State—he grew up in Cornwall, New York, and I was living at the time in Greenwood Lake, New York—after high school our paths did cross on and off. We played different sports. He was a soccer player, I was a basketball player. But what I remember about him was he is who he is today. I mean, he honestly hasn't changed. He's very focused, very driven, determined, bright.

At what point in your childhood did you begin reading?

Wow, that's a great question. Well, on my own of course I started reading, I suppose, when I was very young. If you talk about reading in the context of the profession, I had no interest at all in military history or the military profession until I went to West Point. I was determined to go to Manhattan College in New York City and probably as a pre-law student, so I was always kind of inclined toward the humanities. Then I went into the engineering school. I started reading voraciously, really, at West Point. I've always believed that reading is an important part of who I am. Not that I read a book hoping to gain some discreet insight out of it, but rather to find something that I can tie to something else.

Is it possible that you could remember some of those that were most influential on you early on?

I think so. Reflecting on the answer to the previous question, I read Shakespeare, for example, avidly. I went back to West Point as an instructor in the English department and one day the department head held up two books in front of the new instructors, and he said, "This is the dictionary. This'll give you the definition of words, the words of the English language." He was holding in the other hand the complete works of Shakespeare. "This is where you'll go to find out what they mean." I was kind of captured by that idea, that there's the definition of words and then there's their application. In the '70s I was reading books about the aftermath of Vietnam. I don't remember any of the works in particular, to my discredit. I was reading a lot of the things coming out of TRADOC, about how we change our Army paradigm from one that was focused exclusively on counterinsurgency in Vietnam into something that became Army doctrine. But I was reading avidly. When I went to grad school and then back to West Point on the faculty, I became interested in the profession. If you remember, that period of time in the late '70s was a transitional and transformational time from the conscript Army to the volunteer Army. We were examining our profession, and I think Samuel Huntington's, *The Soldier and the State*, was at that time the primary work on what it means to be a member of the military profession. I was interested in the profession of arms. A fictional work that was kind of linked to those was, of course, Anton Myrer's *Once an Eagle*. And then in terms of getting beyond the foundational elements of the a profession, I became interested in T. R. Fehrenbach's *This Kind of War* which was actually subtitled *A Study in Unpreparedness*. So as I left the West Point faculty and began to migrate back into the force as an S-3 and XO, I was kind of captured by what it means to be a professional now, because I had made that career commitment, as well as this notion that America generally goes to war and finds itself ill-designed, ill-prepared for its first battle.

This is going to sound like a strange question, but does the ghost of General DePuy show up every now and then in the hallways and corridors at TRADOC?

Not only the ghost of General DePuy, but some of his protégés show up in the hallways in a very positive way. I mean, I've got a very strong partner in General Donn Starry, who lives up the road in Williamsburg, and we exchange letters and e-mails. If we're getting ready to do something on something I know he was deeply passionate about when he was the TRADOC commander, whether it's a new learning model, or leader development, I'll send it, I'll shoot it to him, and I'll send him a one-page précis on something we're thinking about and he'll send me back 15 pages of thoughts. It's fascinating. Also General Paul Gorman, who really did a lot of the heavy lifting to make General DePuy's vision happen, and Bill Richardson, also a former TRADOC commander. But sure, the shadow of General DePuy, if not his specter, is always there.

I noticed that you were conspicuously absent from the Pentagon for most of your career.

Yeah, only very late, really, as a colonel I finally made my way into the Pentagon.

How were you able to do that?

Well, it wasn't intentional. You'll hear some folks say that they avoided the Pentagon. I didn't avoid it, it just avoided me, I suppose. There are windows of opportunity and I missed one of the windows by staying in Germany for a very long time as a major, in fact, five years. Into my early lieutenant colonel years I was doing things at the tactical level. Then I came back to be the Army branch chief in Alexandria, Virginia, then on to the National War College. I did come to the Pentagon to work for General Casey in the J-5 and then eventually up for the chairman of the Joint Chiefs as his special assistant.

The 3rd Armored Cavalry Regiment, as you know very well, is the last one we have.

It is.

I know that this unit dates back to the 1840s. When you were the regimental commander, were there certain traditions that were built into that position that made your job easier or more challenging?

I've always believed that the history of the unit is a huge advantage if you choose to leverage it. For example, I'm the 67th colonel of the regiment. In fact, if you ask my aide, ask him my name, he'll say, "Who do you mean, the 67th?" because he was with me in the 3rd Cav. If I go back to the 3rd Cav., when I walk into the room they won't say ladies and gentlemen, the TRADOC Commander, they'll say ladies and gentlemen, the 67th colonel of the regiment. Patton was the 28th. You know this history and the history of the Sergeants Major of the regiment, as well. Those kinds of histories are powerful and we use them to our advantage. It's important to the people, to make them feel part of the unit from the beginning and hopefully that they would feel a sense of belonging.

You've spent much time with troops. Do you believe that your leadership style was forged earlier than a lot of folks who didn't have as much face time with troops?

We're actually examining that balance of tactical and operational strategy. Because, as you know, in a career—I just went over 36 years, so what I'm about to say sounds almost ludicrous—but time is the scarcest resource in a career. That's especially true in the company-grade years and those early field-grade years but the same applies for non-commissioned officers. If you were to ask the question a little differently—when do I think I probably had the leadership skills and attributes and characteristics and qualities that I have today—I'd probably suggest to you that I think in my platoon-leader years, in my troop-command years. I was in the 10th Cav. in those days, the troop-command years. My leadership style was still adapting, changing, I was revising it. Sometimes I think deliberately, sometimes probably not deliberately. Sometimes it was just happening. I think, as I look back on it, I am the leader I am today and that leader is the same leader that I knew when I was a battalion XO. So, that came in roughly 1988, 14 years before I made general officer. I think it was a combination of the accumulation of experiences to that time, but also importantly the time that I had at Duke University and at West Point to reflect on. I really do think that it is the opportunity to reflect on experience that actually forges something into the hard metal that doesn't change. I think you have to reflect and not just be in situ in order to finally determine the leader you will be.

I don't know who said it, but I have always thought that it was an apt comment, and that is that leaders are born because people want to be led. Do you believe that?

We could exchange clichés about leadership almost all day, because it is a lifelong avocation of mine, the study of leadership, and that's why I've been thrilled to have this job. I don't know if it's because people want to be led or have to be led—there's probably a nuance there that's important—but I do think a related point is to what extent you are born with certain characteristics, like the willingness to listen. Is that something you're born with? In my view that's a very important leadership trait, particularly the more senior you become. It's a rare senior leader who is said to be a good listener, so it seems to me to be an important characteristic. To roll back on your point, is that something you just have a propensity to do, or are you born with it, or can you come to appreciate it over the length of your career? Then the other factor of leadership is the one of experience. When I was a captain in the 10th Cav., commanding B Troop, I was absolutely convinced I could have commanded the squadron but for the opportunity to do so. When I became a battalion commander, when I became a regimental commander, I realized that I developed just by virtue of time and experience and reflection. I developed some instincts that probably can't be quantified. I think people become leaders due to both timing and opportunity and it's a choice. I mean, nobody presses you. I say that when I go to Leavenworth to talk to every rising battalion and brigade commander in the Army. First thing I say to them is, was anyone pressed into service? Has anyone forced you to take command? They all look at me like I've been hitting the sauce early in the morning, but my point to them is okay then, you've made a choice. You have volunteered to be a leader. Thank you for doing that, but here's what comes with that, here's the responsibility that comes with the assumption of command.

During those early years—14 years—at what point were you really impressed with the leadership of your peers, your superiors, perhaps even one of the NCOs? How early in your career was it obvious that the Army had good leaders?

Almost immediately. I reported to 2nd Cavalry in January of '75 and I reported to the squadron XO, a guy named Bill Crouch, who eventually became the Vice Chief of Staff of the Army. Then I reported to my squadron commander, Lieutenant Colonel Nick Krawciw, a Ukrainian immigrant who eventually became commander of the 3rd Infantry Division and retired as a major general. My troop commander was a guy named Jim Knowles, who became a brigadier general and assistant commandant at the Armor Center at Fort Knox. The battery commander in the squadron was Tommy Franks. It wasn't apparent to me at the time but looking back on it the 2nd Cavalry was considered to be an elite unit at a time when the rest of the Army wasn't a very consistent organization. We had units that were organized, manned, trained and equipped differently. We happened to be on the border mission on the East German/Czechoslovakian border so we had a higher priority. People migrated there because it was a meaningful mission, and then it was cavalry, you know; there was always that bit of panache. So early on, funny enough, it wasn't readily apparent to me that they were going to be who they eventually became.

I really loved my first troop commander, Jim Knowles. I think that's always the case. It's like your first drill sergeant if you're a soldier. We've kept in touch with each other through the years. But I think the guy that probably had the most influence on me, and the leader I thought I wanted to be, was the one that never made general officer. His name was Jim McQueen. Interestingly, he was the commander of the 10th Cav. at Fort Carson, Colorado from roughly 1979 to 1982. When I went off to Leavenworth years later I wrote an essay for Military Review Magazine about leadership using him as the caricature. I didn't name him but used him as the image that I was trying to describe and I came in second place in the writing award at Leavenworth that year. The point is, I think, I've been blessed. I could rattle off the names, like my brigade commander in Desert Storm was Rob Goff, who retired as a two-star. But to best answer your question, the first leader who kind of lit my fire—that's a description I sometimes use—was Jim McQueen.

What was there about him that lit your fire?

The essay I wrote was about leadership. In it I said, look, there's a whole bunch of leadership qualities that we describe in our doctrine but let me mention one that I'm not suggesting should be added to the list, but it's an important quality—repose. There's a book by Evan S. Connell about Custer and the Little Bighorn called *Son of the Morning Star*. Some of Custer's peers had said of Custer that he lacked the repose necessary in high command. I got interested in what the word repose meant. You know, it was something about self-confidence, it was something about patience, it was something about the willingness to avoid giving credit to yourself, to give others the credit, or to take the blame. It's about putting your subordinates in front of you. It was all these things and I found that the word repose in a positive way captured Jim. That's what impressed me about him. He was tactically very proficient, very technically proficient in terms of things like gunnery.

He really knew the tank, he knew its capabilities, and he was a skilled trainer. Under him we were doing quarterly training briefs long before they were in vogue and mandatory in the Army. This was coming out of an army in the 1970s that was micromanaged because there were so many problems with it. He was the first one that I'd ever seen who said just let go of the reins and see how your subordinates perform. I saw that as something akin to repose and I was actually quite taken by it—by the trust that he not only talked about but actually exhibited.

What do you remember most about the time that you spent with General Shelton?

General Shelton's such an imposing figure. I really enjoyed it. I was a colonel, and you know, he gave me some projects that scared me. I thought, my God, this project has national security implications. I think it was the way he interacted with other very senior officers as the chairman of the Joint Chiefs, and he taught me a lot about interacting with those people. The chairman's authority is somewhat built on collaboration and trust, less on direct lines of authority. For example, the combatant commanders during the period of the Kosovo campaign and others had big personalities like General Wes Clark and the combatant commanders who had a direct line to the Secretary of Defense. The way I saw the chairman use indirect influence rather than seeking direct influence, and the way he worked to build an environment of trust and collaboration, the way he articulated his advice—this sticks with me. I once suggested to him that he should advocate a particular position. He said, "No, no, Marty, you don't understand." He said, "The uniformed military gives advice. We don't advocate." I think there's a fine line there that defines our profession, and I think General Shelton really, really had that sense of what it meant to be a professional military officer in a political environment.

I remember having a discussion about the Eight Imperatives with General Vuono, and he said he thought they were a watershed for the Army. Did you feel that any of those touched you as you were working your way up to colonel?

Any of those imperatives? Yeah, absolutely. Every one of those imperatives, to include the ninth which has just been added. As you know we at TRADOC authored that strategy and took a lot of time to determine which of them rose to the level of being an imperative, but also took a lot of time wordsmithing with the Chief and the senior leaders of the Army to make sure that we articulated them correctly. In reflecting back on my career I think every one of them touched me personally. The one we left out, by the way, just happens to be the non-commissioned officer, soldier as a professional. Don Snyder was an adjunct professor up at West Point for many years. At a conference recently we wrestled with whether we're going to stay at this. I mentioned Huntington's *The Soldier and the State*, the 1980s, examining the profession after the transition from Vietnam and the pressures of eight, nine years of war, an age of transparency, a ubiquitous media, globalization, the security environment which is very competitive now. We question whether we can use words like dominance and supremacy. I think that the world's a little flatter than that now. We've seen the debates about the surge and all kinds of things where the professional ethos has been pulled and tugged and the apolitical nature of the military profession—notably senior officers both active and retired—has been kind of pulled and tugged. It occurred to me that it might be the time to examine the profession again, so we pulled a group together and Don Snyder started it off with exactly the right approach. He said, "Listen, you're not a professional just because you say you're a professional. You're a professional if your client considers you to be a professional." It was brilliant, really, because what he was saying was look, it is time to look at ourselves and determine what and how we are delivering what we must deliver uniquely to the nation, because that's our client. So that is the ninth imperative.

I understand that the four-star community gets together at least once a year.

Once a quarter.

I also understand that one of the things you do is take a look at the one- and two-stars—and especially the three-stars—and as-

sess who's shining and who's not, and where some might be going?

Well, you'd be disappointed if you didn't think we did something like that and the exact process by which we examine and build our bench. This really all falls under the general category of succession planning, doesn't it? Talent management and succession planning. So of course we do. There's a distinction to be made. The one-stars are still selected by a board, the two-stars are still selected by a board, and they both receive officer evaluation reports. I think you know that the two-star rank is the last permanent rank.

Yes, sir.

Everything above that is actually not a promotion, but an appointment. So we look at the board results. We have no influence over the board results for one- and two-stars but we take the board results and then we determine how best to use those general officers. We think about a talent management cycle. We'd say, access: where do you get them? Develop: how do you develop them? Retain: how do you make sure you keep them? And employ: where are you going to use them? If you want to build a corps commander as a three-star then you've got to get him in through division command. This is about taking that which is given to us at one- and two-star rank and employing it for the benefit of the institution and the individual. In the case of a three- and a four-star, where there's no board process per se, we actually find a consensus among the sitting four-stars about who has the greatest potential to be nominated for those nominative positions in the three- and four-star community.

Is it still pretty much a dictum in the Army that you're not going to make general if you haven't commanded a battalion and been a brigade commander?

It certainly depends on branch. We don't talk about combat arms branches anymore; we talk about FSE—Fire, Support, and Effects categories. There's a culling and vetting in talent management too, and so the vetting is the boards that meet for selection for command or its equivalent. There's an O-5 selection list for battalion command, just as you say, but there's also a central selection list. For example, if you're an AG officer, you're centrally selected or not for the G-1 position. So there is an O-5 vetting that you pass through and certainly you have to pass through it if you hope to become competitive for the next level. Then there's an O-6 vetting and then of course the brigadier general board. That said, the important point I want to make is there's no completely proven path. There is a path well traveled but it doesn't mean that there's not the opportunity for someone to travel a slightly different path. On the other hand, the one thing that the military must deliver—the Army in particular does a lot of things for the country—and it can, and it should, and it must deliver lethal effects. That's unique to the military. It's why the military must be a profession, because you don't want a bunch of amateurs delivering lethal weapons. Since that's our foundation, you will generally find that the path to higher rank and positions of responsibility passed through those places where you manage that, where you are the battalion commander of a tank battalion, infantry battalion or field artillery unit. Not exclusively, but generally, and I think you'll find that our Army will always most clearly reward, not exclusively, but most clearly reward that path.

General Alfred M. Gruenther

Although he spent seventeen years as a lieutenant, Alfred Maximilian Gruenther eventually rose to four stars as one of the most distinguished generals in the Army's history. Born on March 3, 1899, in Platte Center, Nebraska, he graduated from West Point in 1917. Prior to World War II his assignments were routine and uneventful. In September, 1941, he was promoted to lieutenant colonel and the following month became Deputy Chief of Staff of the Third Army. The following year he joined General Eisenhower's staff in London, earned promotion to brigadier general, then became chief of staff of Mark Clark's Fifth Army in 1943. After the war Gruenther was named the first director of the joint staff of the Joint Chiefs of Staff. In December 1950 he became chief of staff of SHAPE and, in July 1951, pinned on a fourth star. He succeeded General Ridgway as SACEUR in July 1953 and retired from the Army in November 1956. From 1957 to 1964 General Gruenther was president of the Red Cross. He resided in Washington, DC, until his death on May 30, 1983.

6

"I WAS MORE OF A FOLLOW-IT-UP-YOURSELF MAN"

A conversation with General Alfred M. Gruenther, USA
October 17, 1975

General, I know you like to bring up in your many speaking engagements the fact that you spent seventeen years as a lieutenant. Fortunately World War II came along.

I start off by saying that luck plays, in certain cases, a very important role. Now, my future more or less rotated around Eisenhower. Even when I retired, I became president of the American Red Cross which is just about two blocks away from the White House. So I saw President Eisenhower quite frequently. I was just invited to a party here a week from tomorrow and I'm looking up to see if I knew the fellow, his first name was Brian, but I'd forgotten what rank he was. He's a colonel. He's retiring. He was in the class ahead of ours. He's inviting us to a cocktail party, but he must—in his somber moments—wonder why in the heck did that guy get to be a four-star general and I got to be a colonel. I'm sure, as he analyzes it, he analyzes that he must have been lucky. He knew that guy Eisenhower and that saved him. I just bring that up because when you go enunciating these things, of studying very hard and so forth, you do run into the question of a matter of luck.

Just to give you another thing on the question of luck. Now we who have graduated early have come back and they gave us some special courses. Lo and behold they select me as an instructor after one of these special courses and now who are my students? Well, I'm now 20 years old and I don't have anybody in my classes younger than 27. And what am I instructing in? I'm instructing in things such as elementary law, mess management. I've had nothing to do with mess management. Physical culture? Well, hell, I don't even skip rope well at that stage of the game. So, it was really called the department of miscellaneous subjects. Well, I studied to beat the devil, don't you see. I participated in none of the activities. I'd go into the officer's club, I got married of course, and I passed fellows playing bridge. I more or less spurned them, don't you see. Here are these guys with a career in front of them and they're playing bridge at night while they should be at home studying, because that's where I'm going as soon as I get through. A little while later, we're invited to a dinner at our boss's place, Major Gruber. When the dinner is over, he gets up and goes to the closet over there and he pulls out three card tables. Well, I count the house and there are exactly 12 people there. So I went over, just sauntered over, don't you see, scared to death, and said, "You're going to play cards, Major?" He said, "Yes, we're playing bridge. You play don't you? I said, no, sir, I don't. Well, he said, you're going to play tonight. Well, haven't you ever played cards before?" "Yes, I did." "What kind of cards did you play?" "Well, as a kid I played high five and euchre." He said, "This is just like high five." Of course, it's no more like high five than a pig is like Sunday. So I drew his wife, rotated in some peculiar way, and of course I was just embarrassed to tears.

So that night when I went home—it wasn't too late when we left, about 9:30 or something like that—I went over to the club, pulled up a chair to watch this bridge game. They looked at me and said, "Oh, Mr. Big, what brings you here?" I said, "I just found out that maybe you've got to play a little bridge to succeed here." Well, they said, "You've never come here before." I said, "No that's right. I've just been to dinner and I got embarrassed." Well, I became quite a bridge player in due time, but I bring it up because again that was a matter of luck. But, you see, I was no longer a spring chicken. This is 1940 and I was one year older than the year. It's 1976 now, I'm 77 years. old. This was 1941 and I was 42 years old. Well, that's not a spring chicken anymore. So you have a certain amount of luck coming into this thing.

But isn't it true that if you don't have the tickets to take advantage of that luck, it doesn't mean anything?

That is right. I would say this, using your language, you do have to keep boning all the time to be sure that you've got the tickets. But having the tickets isn't automatically the thing that is going to put you up. You've got to have some luck in assignments where your particular talents work out well. Just to show you one where it worked out well, during this period—when we were ordered back to West Point from December and remained there until June—we took everything, including chemistry. How much time did I have to study chemistry? I think that we had chemistry for that period for three months, the reason being that—having this bunch of kids on their hands—they were trying to do the best they could by them. Apparently, one day, they had a chemistry man, Colonel Robinson, come in, and I recited very well. And apparently he put that down in a book some place. Ten years later, 1928, I get ordered back to West Point as an instructor in chemistry. Well, my God, I almost fainted to get that. So all during the year, I had eight months' notice, I went to night school studying chemistry. And when the summer came—I didn't have to report up there until August—I took an intensive course, but I certainly didn't know a great deal about chemistry. And to wonder how I got selected, it again was a matter of luck. Colonel Robinson, the head of the department, he would bounce in on classes, don't you see. Now by the time I got there, he had passed on so I had no way of going around and saying, would you mind looking at your notes and seeing how you happened to write down a note about Gruenther? He had gone on to heaven and, I hope, is having a very rewarding time. But, it was luck. And I did fairly well.

Walter Krueger went out to the Pacific and became a famous four-star general. Have you ever speculated on whether or not you would have gone as far with him?

Oh, I might very well, because he had promoted me to chief of staff, but you see the Eisenhower thing went even farther than the Krueger thing. Krueger was, by that time, about 65 years old, don't you see. So as soon as the war was over it was out. You could tell that Eisenhower had his future in front of him. So, I get another shot at Eisenhower. Now then, we're up to 1950. In 1950 they're getting ready to start NATO. One night I was with General Eisenhower when Drew Pearson came on the radio. He didn't like Drew Pearson. Drew Pearson came on on a Sunday evening and he predicted that General Eisenhower was going to be the supreme commander of this new command. It was just newspaper speculation then. So, I said, "Well, Drew Pearson knows it, that's nice, congratulations." He said, "It's not only congratulations. If I got hooked for the job, you're going to be hooked also!" Well, that was another way of saying that I was going to be kept on. He didn't like to be kidded about Drew Pearson. He was annoyed that I had the radio turned on, listening to Pearson, but it was in my home so he couldn't do a hell of a lot about it. We hadn't sat down for dinner yet. And Pearson was right, don't you see. About 10 days later Eisenhower was selected, so he called me up and said, "Get busy." Supposing I had gone forward with Krueger; by 1950 Krueger was a thing of the past. He might have even been dead by that time, I don't remember.

So, I'm in favor of the usual things but I'm just bringing up the fact that you do have a case of luck. Again, I got a considerable degree of competence in bridge. Having had this embarrassment with my boss, I then started studying bridge a little bit. A year and a half later I got ordered to the Philippines where we had a little more spare time, so I played quite a bit of bridge there. By the time I came back I knew a little about bridge and became an instructor at West Point. Another fellow and I, when a tournament was announced, we decided to go down to the tournament. The important guy in bridge in those days was Culbertson. We get assigned to Culbertson's table and we're there and the tournament is late starting so I know I'll get all the information I want about bridge. But instead of that, Culbertson kept asking me questions about the military. I never got a chance to ask him a single question. The tournaments were always starting late in the afternoon and later still starting at night. I kept thinking, God, what these people need is a disciplinarian to get things started, so I started running tournaments at West Point to find out what there was to it. Then I invited about 8 or 10 of them to play in an evening game and come to West Point to see a parade and they came. Our game was to start at 8 o'clock and we started exactly at 8 o'clock. Well, they didn't say anything about it, but when it came time to get somebody to run that tournament the next year, I was given the job, don't you see. And they made it clear, from a fellow by the name of Goldman, they made it clear that the reason that I was given the job was not that I was a great bridge player, but that I knew how to start a tournament on time and was disciplined and would not be bulldozed by a lot of hot shots. Well, we get our first tournament and I boned it up getting people to be there on time, and they're all on time except Mr. Goldman and his partner. So I threw him out and brought another pair in. Everybody knew because I kept calling, "Mr. Goldman, Mr. Goldman". This was in a New York hotel and I thought he might be in the restaurant. We had waiters go out and page. No luck. So I said, "Mr. and Mrs. So and So will you please come."

About 15 minutes later Goldman comes in with his partner and says, "Lieutenant, where's my seat?" I said, "I'm awfully sorry. You don't have a seat." He said, "What do you mean I don't have a seat?" Well, I said, "It's 15 minutes late and we had to start the game on time." He said, "I was out on tournament business," which he wasn't. I said, "I'm awfully sorry about that but it's too late." He wouldn't speak to me for six months. It helped me out a great deal, though. Then he got so that he thought it was a hell of a good idea, don't you see, and we became very close. A curious thing. A year and a half ago I'm on a board called the Advertising Council. They've selected a student each year and they wanted me to look up a student in American University. I kept calling his fraternity house, but I could never get him. I'd leave messages, but could never get him. One day I called up and gave my name to this fellow. "Oh," he said, "I know all about you. My name is Goldman. My grandfather is still talking in his grave about you, I'm sure."

General, what about the subject of leadership? What can you tell me about your leadership style and what made you effective in getting things done?

I would say this, curiously enough, in a sense in bridge tournaments you had to have everything in line and then be fortunate to have the people who would come in and sit down at the proper place at the proper time. You had to do so much coordinating ahead of time that, curiously enough, that helped me out a great deal in seeing that the chips were in the right place. In effect, that is it. Either you're doing it or you're having somebody who will come and report to you that such and such a thing is the case. Then, throughout my career I had a chance to put those principles into practice quite a bit. It helped me be a fairly good manager. I missed a lot of things too. You make a lot of errors. The question is whether you can profit from your errors and that's one of the things that is always a problem. I was more of a follow-it-up-yourself man. For example, I had the habit of writing notes—I don't think I have any more of them here anymore—but I had a little, narrow notepad, especially over at SHAPE. They were called "Gruenthergrams." Going through the hall and seeing that the clock wasn't on time, I'd write a note to Captain Smith, the clock is four minutes behind time. Please see that, at least, this clock is kept right. It was a case that illustrated a point that I was following up on these things. Now when it came to the case of Field Marshal Montgomery, who died just a few months ago as you know, here was a man senior to me and I was his boss. Well, I didn't follow up on him. I consulted him very much and came out very well with him. He admired me and I admired him. But it was a tricky situation to have him under me.

When you wrote these notes, how did you know what kind of time parameters to put on and how did you know to follow up with a certain person with a certain problem?

Well, they didn't fiddle around with me because if I had asked a fellow to get a report in by next Monday on a certain subject, I would remember that I had done that. I had a fairly good memory. Having had somebody miss the boat and then catch hell about it, he spread it around that if General Gruenther tells you to have it in by 8 o'clock Monday morning, get it in at a quarter of 8 at the very latest. He'll be there and be checking on you.

So you put a great deal of stock on accountability for actions.

Yeah. But now then take General Eisenhower. He wouldn't have done that. The reason why I succeeded with him in previous things is because any time he would be doing this I would be the guy following up on it and checking and checking and rechecking. That made me almost indispensable as far as he was concerned. I'd follow up on little things. He had somebody in his office one day and he used a word which was mispronounced. I was pretty sure it was mispronounced. I can't think of the word right now, but it's one that we use quite a bit. On the way out I stepped over to the dictionary. He had a dictionary in his office. He said, "Don't bother about it, you'll find out it's pronounced both ways."

He knew just what you were going to do.

Yeah, but he was wrong. I didn't dare look at it then but you can rest assured that when I got back to my office I looked at it and found out. A little bit later I told him.

You didn't send him a Gruenthergram?

No. (Laughter)

General Thomas T. Handy

Thomas Troy Handy was born on March 11, 1892, in Spring City, Tennessee. Commissioned in the field artillery in 1916, following graduation from VMI in 1914, he served with the 5th Field Artillery, 42nd (Rainbow) Division and 151st Field Artillery in France during World War I. Handy served with the Army General Staff in Washington 1936-1940 and 1942-1945, rising from major to four-star general in that period of time. Following the war he commanded the 4th Army, the European Command from 1949-1952, and he was Deputy Supreme Allied Commander in Europe in 1952-1953. General Handy retired from the Army in February 1954 and was active in the private sector for many years thereafter. He died in San Antonio on April 17, 1982.

7

"YOU'VE GOT TO GET OUT AND SEE THINGS YOURSELF"

A conversation with General Thomas T. Handy, USA
May 1, 1975

General Handy, the first question I'd like to ask you, sir, is what made you successful in the army?

I haven't any idea. I think I was very fortunate to tell you the truth.

We were just talking here a minute ago and you said you thought that a lot of success in the military simply has to do with being lucky.

That's right. I think a lot of these things are controlled by fortune or luck or whatever you want to call it. The only thing you can do yourself is to be ready for these things, or attempt to be. Whether you get the opportunity or not, I think a lot of it is a matter of fortune or luck. As a matter of fact, I think it applies all the way through. I think people with big reputations, which they undoubtedly deserve, they've often just been very lucky. Like these so-called high-powered military strategists. Well, if you happen to be right, and nobody's wise enough to know ahead of time whether you're going to be right or not, why, you're a great guy and very wise. On the other hand, if you happen to be wrong, and many times the wrong isn't that fellow's fault at all, well you're out of luck. Your name is mud.

In your case did you feel for the most part that you were at the right place at the right time?

Yeah, I think I was very fortunate, very fortunate in many ways.

Well, talent has to play a factor, doesn't it, General?

A fellow has to have some ability, he can't be a complete dumbbell, there's no question about that. I'm assuming that he has some ability. And a great many things affect it. Two things I think are important always. One of them is the job being more important to the fellow than himself. In other words, I think all those people were devoted to doing their jobs. Of course a lot of people are ambitious, some more than others, and that's all right, too, but I think the job's a little bit bigger than the man. I mean in his mind. I think that's extremely important. Old Hap Arnold, did you ever know him?

No, sir, he was a little before my time.

Well, he was quite a remarkable man. I remember one time he said something that made quite an impression on me. He said that when he was picking a man for a job, a key job, an important job, what he looked at was not so much how well the man was doing the job that he had right then—he looked at that, too—but whether he was the fellow that had grown with his job. I think there's a lot in that. People often times reach a certain level. In athletics a fellow will be a wonderful star in the minor leagues. Well, he goes up to the big leagues and he's a flop, don't you see, and the reverse is true. The fellow may not have been absolute in the minor leagues but as he goes higher he gets better. In other words, he grows with his job. I think that is important. I think, certainly, most of those people, the ones I know, did.

That's a very interesting point. I think, now that I reflect on people in the military, you have to be able to do that or you couldn't be successful.

That's right. Certainly, you couldn't. It depends upon what you mean by success, but other people more than us, more than the Americans in the military, accept the fact that man reaches a level. Way back there in the French army, a fellow gets to be a captain and stays there until he's 50 years old. He's reached his level, don't you see. He knew it and they recognized it and let him stay on. Well, in ours it's different. But people do reach a level. A guy will be a hell of a good captain or a major or a colonel, maybe, and when he gets up higher, he won't be so hot. And the opposite is true, I think. What old Hap Arnold said, I think, was absolutely true. It's very important whether a man grows with his job.

General Marshall was also a graduate of VMI, a few years ahead of you. Did your paths cross much in your later careers?

Well, of course, I have to always do this. I'm asked a lot of questions about General Marshall because I worked for him pretty closely for quite a while during the war. What I say about General Marshall should be qualified by the fact that I'm, in his case, a very prejudiced observer. For my money, I've seen a good many of these big shots. General Marshall is in a class by himself. I don't think anybody that I've ever known was in the same bracket. He was a very, very remarkable man in many ways.

I never knew General Marshall until I was back there just before the war in the War Plans Division and he came in there as head of that division. But, as a matter of fact, he didn't spend much time with us because he was working over with the Deputy Chief of Staff. I think they were getting him ready to be Deputy Chief of Staff. This wasn't too long before the war. He was the head of the division but I was way down the line, very junior. Of course the divisions weren't as big in those days as they are now. Even in the same class I think we had a total, when I went to that division, of about 11 officers on duty including the head of that division.

In the whole division?

The general staff, you see. Maybe it's better if there aren't so many.

Of course my tour expired in 1940, and I left Washington and went down to Benning. I was down there for a year with the 2nd Armored Division and then I was yanked back. I wasn't supposed to go back. When I went back in '41, the summer of '41, he was Chief of Staff then. I worked in War Plans, and Gerow was head of it when I came back. I came up after Pearl Harbor and he headed it, then when Ike left I was put in there as operations. The old man used operations, as he expressed it, as kind of command post so I saw quite a bit of him.

When it comes to leadership, General, what did you and the members of the General Staff in Washington think about the Germans during the war?

The fact remains that for quite a long period—it wasn't a flash in the pan—they were uniformly successful. Isn't any doubt about it. Every campaign they had was not only a success but far beyond what anybody could reasonably expect or we thought would happen. The Germans are awful good soldiers. Sometimes they miss the biggest point of all, as you know, but they are awful good. Now you take this—I've often said, to illustrate about the Germans, they always accuse us of trying to fight the last war. Well, hell, I don't know what else you do but study what's gone on before. Everybody else does—an engineer or a doctor or an author or anything else. It seems to be with the military they count it a crime to study what's gone on before. But everybody has to start that way. Now let me tell you something. The Germans were the only guys I know of who were really smart enough to really dope out a lot of these critical questions even before the war. You take this use of armor. There isn't any question that the Germans showed the way on it. As you know very well, the people who invented the tank and used it first were the British back in World War I. We played with them for 20 years and so did the British and wrote books and everything else on it. The only people that really got down and doped out how to use tanks before the war were the Germans, and they did it. And they showed everybody how to do it. A lot of these people have studied weapons. They talk about new weapons. Well, hell, somebody said a new weapon is not the important thing. They're important right enough, but it's to find out how to use it. If you go back through history you'll find out—and this armor is a good example of it—that it was years after a new weapon was developed before people found out how to use it. The Germans were able to do that, so you've got to take off your hat to them.

When Ike began getting the people together for the invasion of Europe, he said that you've got to get rid of anybody that you don't have explicit faith in doing the job. From your vantage point in Washington, General Handy, did you see any people who were capable, but for one reason or another ended up not getting any further, because of circumstances that might not have been in their control?

Of course that's a very difficult question because you're getting into the "if" thing—what might have happened. Nobody knows. A guy might have been a whiz if he'd had the chance and didn't have it. As I say, a lot of it is luck. But this thing of command ability and leadership ability that you're talking about that Ike was trying to emphasize is probably the most important thing of all from a military viewpoint and maybe some other viewpoints—national—too. I think what he meant was—and maybe what you're driving at—it wasn't the people who didn't get there but the people who got there and then were relieved. That is, I think, the most difficult thing that any commander has to do—because he relieves this fellow, it's not because the man isn't a hell of a good man, quite capable and also working his heart out—but he's just not getting the job done. It's one of those things that has to be done and if the fellow can't do it you've got to get somebody else to do it.

General Marshall always backed up these people in the theaters, in the field, on these reliefs. Some of the people relieved were personal friends of his and had been for years. But I think I figured it this way: you put a guy in charge and tell him you've got a certain job to do. Well, he has a subordinate in a key position there that he feels can't or won't do the job for any reason at all and he says, "I need somebody else." Well, I think you've got to more or less be guided by what he says or, if you've lost that much confidence in him, you'd better get somebody else to run the job. It's that kind of a thing. It's not one of these things about the fellow's guilt or innocence. It isn't that kind of a thing at all. These people that got up, say, to division command, they were damned good or they wouldn't have gotten there. They had a hell of a fine record and everything else. They had a lot of ability and everything but they just didn't cut it when the wool got awful short. It wasn't their fault at all and it wasn't because they weren't damned good men and they weren't working their hearts out. As a matter of fact, sometimes—and I've told this story before without mentioning any names—but the fact that the fellow is such a hell of a fine person, really fine person, sometimes can work against him. I know of a case of a division commander who was a hell of a good friend of mine—still is—who was relieved. I think the main reason was that he was such a hell of a fine person. That fellow was the kind of a man if he had a man killed that was in his division it was almost like losing his own son. Well, a guy can't command a division in battle feeling like that, as you can very well imagine.

Look at World War I and see how many people we had relieved. General Pershing knocked them off just like that. It was a really remarkable performance. Mr. Churchill came to Washington after the war and talked at the Pentagon one time. He said that he's never had any doubts about America industrially. He knew that it would take a little while but eventually they would get the stuff, they could do it and all that. He said the thing that was absolutely amazing to him, and he still didn't understand it, was how we, without any army practically for many years, developed and exercised the high degree of leadership that we showed in World War II. He said that was something he still couldn't understand. So I thought that was quite a compliment.

Right, that's exactly what I meant.

It's hell, this business of relieving people, because I say, it isn't because the guy's done something wrong or he's lazy or anything like that. Hell, he's working his heart out. He's a hell of a good man, yet you relieve him. It's not easy.

General, when you were in charge of all American forces in Europe after the war how did you find out what was going on in the lower ranks?

You've got to get out and see things yourself. I think that is absolutely necessary and you get a feel. You get all these reports and everything on some outfit or something that's going on but you get a different impression when you get on the ground and talk to people up and down the line. I don't know, I think that every guy has a different way. The first thing that he has to be is himself. The worst mistake that a lot of people can make is trying to be like somebody else. I saw it back in World War II. Hell, some of our best people couldn't be like George Patton and neither could he be like them. He was Patton and Bradley was Bradley, and they had different ways of doing things. Both of them were very effective, wonderful leaders, but when you say how do you do it—I don't know.

General Handy, how did you go about handling difficult objectives?

I don't know. I think that's one of those questions that I'm afraid there isn't any definite answer to it. It depends on the situations, what you knew already, what had gone before, what the object was, a lot of other things. So, I'd answer with an answer that's as general as your question and doesn't tell you anything. It depends on the situation and probably every one would be handled differently. Now, with a new job, naturally you want to find out all about the job that you can. What's gone on before and what is going on and find out the people you've got. That is extremely important, probably more important than anything else, because if you have good people, you get a good job done. It was like when I took over the Operations Division from Ike. My main feeling was I was just scared to death. All I can say is go ahead and do the best you can. What that best is—it's awful hard to set a bunch of rules or procedures. Maybe the worst thing you could do is to have a bunch of rules and procedures because whatever you make them up for, the situation that's going to confront you is going to be different. Be sure of that.

Did the Army encourage or discourage individuality among its leaders?

I think it did encourage it, very much so as a matter of fact. At

Leavenworth they gave you a lot of people, though—too many decisions and problems. When you were a general's aide you made the decisions. But the Army was rather free in their thoughts and maybe sometimes too free in expressing them. I know when I went to the Naval War College that's one thing that struck me. The Navy always weighed the thought of the senior. He might not be the boss of the problem or the maneuver, they detailed people for various things. I asked some of my friends up there after I got to know them pretty well. We were working on a problem and we were staff and said, how about this, it looks like this. Nobody was saying anything. I said, "What's the matter?" They said, "Let's see what the captain thinks." There was one captain in the group. He wasn't even the head of the group for this thing. I said, "Well, listen, after all it was a theoretical school problem." This is certainly different from the Army. They don't get to express their views at all.

In any discussion of leadership one question always arises—are the truly outstanding and effective leaders born or can a Joe Smith be taught to be a George Marshall?

Well, I don't know the answer to that, to be perfectly frank with you. I think you can improve the fellow and his qualities so that he would be a better leader, but whether you can make him out of whole cloth I have very great doubts. The people that I've known were real leaders regardless of their background and preparation and everything else—and I don't care whether it was a company or a battery or a whole theater—whose personality was felt down through the command. Every one of them was the same way. Every guy in the Third Army, in his own mind, knew something about George Patton, knew who he was and everything else. That's always true. I remember old man Summerall, 5th Corps, way back in World War I. His personality was felt all the way down through it. Most people never saw General Pershing but they knew General Pershing was up there. These people's personalities are felt all the way down through—MacArthur and all these guys, Eisenhower, all of them. Now, they don't all do it the same way. They have far different methods. What works for one man won't work for another. Above all I think the fellow's got to be himself.

There's one fellow you can't fool and that's a soldier, regardless. They're pretty wise. They size up people pretty well.

When members of your staff or certain of your subordinates put forth a superb effort, clearly well beyond what was expected of them, how would you go about rewarding them for that extra effort?

That would all depend upon the situation. As a matter of fact, I was always very fortunate in the people I had working for me. That's the only reason I got as far as I did. They carried me along. They were doing remarkable things, I thought, all the time. You do the best you can. Sometimes you can give them promotions, sometimes you can give them this or that or the other. But I think the guy knows whether he's doing a good job. The most remarkable people I ever had, I think, is when I had Operations there at the War Department during the war. Boy, those guys, they were working 17, 18 hours a day and putting out first class work, which is extremely difficult.

Admiral Ronald J. Hays

A member of the class of 1950 at the U.S. Naval Academy, Ronald Jackson Hays was born on August 19, 1928, in Urania, Louisiana. A naval aviator, he served as a test pilot 1956-1959 and flew numerous missions in Southeast Asia during the Vietnam War. He balanced assignments in the Pentagon with command at sea, including Carrier Group Four in 1974-1975. He was Deputy Commander in Chief of the Atlantic Fleet 1978-1980, Commander in Chief U.S. Naval Forces in Europe and Commander in Chief Pacific, the position from which he retired. Admiral Hays resides in Honolulu

8

"YOU GET THE RIGHT MAN IN THE JOB QUICKLY"

A conversation with Admiral Ronald J. Hays, USN
August 22, 1986

Going back almost to the time of the Civil War, it seems that in the evolution of the Navy and the Army there have been great captains that have sprung up every generation, or every other generation, and those who have had the good fortune to work with and for those individuals were better prepared when they, too, succeeded to very high command, very high rank. Then they became the great captains and so forth. Have we, 40 years after the end of World War II, run out of those individuals?

Those individuals in World War II who could be used as standards to live by, they're all gone. None of them are left. Vessey was the last one. He was an individual who proved himself on the battlefield, an individual who had a feeling for people, a great consideration for those who he came in contact with. He was an example to live by. But he's gone, he was the last one. We've had other wars; Korea, Vietnam. And there are some people who have performed well. John Wickham is a good example, the current Chief of Staff of the Army, who was a commander in Vietnam, badly wounded, did some things that military people look to and admire. I think that is important because exposure to combat can often times reveal either exceptional strength or weakness.

The media these days likes to speculate that all the warriors are gone, nothing but bureaucrats remain and other managers, the feely-touchy people, and if we have a war they're going to be incompetent. That might be true in selected situations. But, it is not true as a general statement, that's my thesis. But we have situations developing that allow an individual to demonstrate the kinds of things we're talking about here; courage, determination, bravery, what have you. We have situations that develop even in peacetime that will reveal in the individual the wherewithal to persevere. Some prevail, some of them stand out. I'm talking now about the good managers.

To succeed in a peacetime environment you've got to be a good manager, you've got to be a good leader, you've got to understand people, because we spend a lot of time processing paper. But the successful senior military leader has got to have a capacity to grow and adapt, like shuffling papers and being "the manager" and being the military leader, the combat leader. If you don't have combat, you can only speculate who has what it takes to successfully lead. Sometimes you make a mistake. You know we can't have combat every now and then just to find out who our best leaders are

In your opinion, are these people going to rise to the top with or without combat experience, with or without any direct leadership in the field, leadership on the high seas? As time marches on, we move further away from people who had that hands-on combat experience. Is that going to be a factor, do you feel?

I think it will be. You see me confronted with a tense situation, a life-threatening situation and then I'm going to respond a certain way. It's either the right way or the wrong way. If I respond in a way that suggests that there's this ingredient you need in a combat leader, and if I have demonstrated—I don't want to use myself as an example, this is the editorial "I"—if I do the paperwork and get the reports in on time, and do the analysis and come up with coherent recommendations, I'm able to converse with the Secretary of Defense or what have you, I'm going to go to the top if I do all those things right.

Now in so doing, if I've never been in combat, I may or may not be the right guy to lead troops, aviators, sailors in combat. And that's an unknown until I'm confronted with it. And once I'm confronted with it the answer will come out.

Now, as you know, in view of your interests and study, people such as I, as I've described, had risen to the top in '39, '40, and '41, and we found, as Nimitz did in the Pacific Theater, that they were his dear friends, they had done a lot of things right over the years, but they were just not right for the combat situation. And we'll have the same experience now because we are unable to identify that individual with an absolute degree of certainty. But the system is automatically cocked to do that. And it takes about that long (snaps fingers) to do that. And it takes about that long (snaps fingers) to replace me. If I'm not the right man to fight in the Pacific, I'll be gone before you can draw the next breath, almost. I'll leave the place.

How do you feel about that?

That's the way it's got to be. Absolutely. You don't have time to experiment, in my judgment, in modern-day combat. You get the right man in the job quickly.

Admiral, to what degree has the rapid change in technology affected your capability to lead and to manage as a senior naval officer?

It's had a profound impact. And that impact is based principally on the availability of information now. It was denied the commander in days past. With the man in control of the arrangements we can now do a far better job of remaining aware of the status on the battlefield and take the appropriate actions depending on what we know and the conditions around you. Warfare has become far more complex. It has become far more deadly. You've got to be a lot smarter about it or you're going to be killed. You can control it better and all this comes from technology.

Admiral Stump, the gruff and combative gentleman who was the first officer to hold the position you now fill—CINCPAC—if he was granted the time to spend a few weeks with you out at Pearl

Harbor now, how do you think he would react?

I never met Admiral Stump, but I have the impression on the basis of what I've read and heard about him that he would not be comfortable now because the leadership styles have changed rather dramatically since Admiral Stump was CINCPAC. Admiral Stump was an autocrat who made decisions quickly, emphatically. Today you find with this information flow that is available that one has to consider, ought to consider the various ramifications of the situation before a decision is rendered. I don't know if you've gotten the message.

Yes, I did, I certainly did. Looking back on your own naval career, has your style of leadership changed significantly from your days as a junior officer? Your basic style of how to get along, how to lead?

I want to say no, I don't think it has. I find that one of the toughest questions that anybody can ask me is, what is your leadership style? How do you lead people? And I don't have a good answer. I know that I feel very strongly and hope that I convey to all around me that I want good results, I want a quality product and I'm willing to work myself or with anybody to whatever length of time and effort is necessary in order to produce that quality product. I believe that you need to be sensitive to the feelings of others. You need to get along with people, all kinds of people, in all strata in the organization. Show some interest. Be willing to put in as much effort any anybody else. Know as much as you can about whatever subject is in the fore at the time, work at it.

Admiral, looking back on your career, what factor was the most important in your rise to the top?

I believe that my success and upward mobility can be attributed in a large part to my combat experience. Had I not had the combat experience that I did, I probably wouldn't be here in this capacity.

In that it gave you the training or gave you the exposure?

I think it probably gave me the exposure. I was in combat at a time when we were introducing new airplanes, the A-6. We had a very tough assignment. A lot of losses, a new system. Never before did we have the capability to go in low-level at night, in horrible weather, in rain and so forth and strike and return. So there was an awful lot of public interest. When the newspapermen came out when we were on station, invariably they would trot them down to my ward room. I was the feature event. I didn't particularly like it at the time, but they always came down to talk to the A-6 pilots. As a result, as the Brits used to say, I was mentioned in dispatches from time to time, and thinking back on it rather objectively, I think probably that recognition caused me to continue the upward mobility.

General Harold K. Johnson

A survivor of the infamous Bataan Death March in the early months of World War II, Harold Keith Johnson was born on February 22, 1912, in Bowesmont, North Dakota. Routine assignments followed his graduation from the Military Academy in 1933 and by 1940 he was a member of the 57th Infantry, or "Philippine Scouts," which were taken prisoner after the fall of Bataan. Following more than three years as a POW, Johnson was repatriated and, returning to the United States, attended the Command and General Staff College. He saw almost constant combat in the Korean War as a battalion commander, then several years of staff duty in Washington. Promoted to brigadier general in 1956, he was serving as commandant of the Command and Staff College when ordered to return to Washington as the three-star Deputy Chief of Staff for Military Operations. A year later he was appointed Chief of Staff, a position he held through the years of heightening involvement in Vietnam. For many years following retirement in 1968 Johnson was involved in patriotic endeavors. He died in Washington, DC, on September 24, 1983.

9

"I WALKED OUT OF THE PENTAGON MY OWN MAN"

A conversation with General Harold K. Johnson, USA
June 4, 1976

Was it Secretary McNamara who was the impetus at the time behind your selection as Chief of Staff or was it General Bus Wheeler?

I don't know. This is one of my very deep regrets that I simply have never talked to General Wheeler in any depth about it. Of course, now that's not possible. But I would say that it was not Mr. McNamara. Mr. McNamara did not name me for that job. Obviously, he had to put a chop on it. Whether it was his recommendation or Mr. Vance's—he was then the Deputy Secretary of Defense—it then went to the President for his approval. I don't know. But I would say again, here you come back to the fortuitous circumstance. I came here to Washington from Ft. Leavenworth in the summer of 1962 on a temporary assignment to conduct one of Mr. McNamara's long list of studies. This one happened to be number 23, called Project 23, which was a study of tactical nuclear weapons on the battlefield. And I struggled with it. I think it's probably one of the toughest intellectual problems that I've ever addressed and I didn't solve it, not satisfactorily. But during that summer I was here alone for most of the time. In the summer of '62 since my wife was not here, I ate in the Secretary's mess. Mr. Vance became Secretary of the Army on the first of July 1962. He and his undersecretary, and many times assistant secretaries, had breakfast there. So I had breakfast with one, two, three or four of them for a period of certainly the summer months.

There were casual conversations about a whole variety of subjects and I, of course, I was very proud of the Command and General Staff College and the things that it did and there was not that much knowledge about it there particularly among the civilian secretariat. So there was an opportunity for them to know me and for me to know them. Now, the next year, in early 1963, essentially on the recommendation of the man who was then Vice Chief of the Army, General Hamlett, I came in here at the end of February to become the Deputy Chief of Staff for Military Operations when General Parker left. That was my first continuing association with General Wheeler, the Chief of Staff.

Then another circumstance occurred. It was a part of the pattern. General Hamlett had a severe heart attack in March and he, I think without question, would have been the Chief of Staff had he been in good health. I don't think he retired until the first of September, but his application was being processed, so he was still under medical care and he was no longer serving actively. After he had his heart attack General Wheeler brought General Harris up here from the Continental Army Command to serve as acting Chief of Staff. When I became Chief of Staff, General Harris returned to Continental Army Command. Now both of us were designated at the same time. I for Chief of Staff and General Abrams to Vice Chief of Staff. He was serving in Europe.

You were picked over 40-some senior officers. How many of those gentlemen were qualified to be Chief?

In the first place, you have to define the qualifications. Not one of the qualifications has to bear upon one's age. There were quite a number of people serving in the four-star capacity. You'll recall that General Wheeler had been in the job just a year and nine months, so that under the normal course of events he had another year and three months to go. You had a number of people who were 57, 58, in the four-star rank who wouldn't complete a normal four year tour by the time that they were 60. Not that there was a mandatory requirement to retire at that time, it was just a custom to retire at 60. Then personalities enter in. How do people work with each other and that sort of thing? Well, without naming any names, I think that there were a couple of antipathies that existed. I go back to what I felt was one of the features that led to General Wheeler recommending me, because I'm sure he recommended me first. I don't think anybody had gone past him because, first, he was Chief of Staff of the Army and, second, his relationship with the Secretary of the Army, the Secretary of Defense and the Deputy Secretary of Defense was such. But his recommendation and his views would have been given an enormous consideration as to who was recommended.

Among the four-star people at that time there were few by virtue of age or some other reason who were logical to succeed. There are cases of not having had all that much time in the Pentagon, and there was a view that you really needed to know how the Joint Chiefs of Staff functioned because this was a primary role. How did you fit in with the rest of the things that were going on at that time? So all of these elements, having nothing really to do with professional qualifications, were part of the mix. I guess that you could draw a relationship between outlook and professional qualifications, I think in a relatively limited sense. I think there's that difference.

Now where did I come in there, coming back to this fortuitous assignment? I was Deputy Chief of Staff of Military Operations, had an enormous number of things under active consideration by the Joint Chiefs of Staff, and I sat at General Wheeler's left hand. I was not the kind who pushed in—"you had to do this or the Army's going to fall apart, had to do that or it was going to fall apart"—because the Joint Chiefs of Staffs' activity is essentially one of striking compromises. What can you give in return for a little take here and there? Not in the sense of trading off, blatant trading, but in the sense of really what's best overall in contrast to what's best for the individual service. I might add as Chief of Staff of the Army I frequently read a lecture to the Army staff at Joint Chiefs of Staff briefings. I said, "Look, we don't have an Army position. I'm independent as a member of the Joint Chiefs of Staff. I use the Army staff because I've got confidence in them to back me up, more confidence than I have in joint staff. What our viewpoint is in every issue that we look at—our first question is—is it good for the country? Because if it's good for the country, it'll be good for the Army. That's got to be our philosophy."

Well, I hadn't quite coalesced that view as the Deputy Chief of Staff for Military Operations. What did that mean to General Wheeler? It meant that he was comfortable, because I could give him pros and cons. You have to do this, you have to do that. So this meant that we had a growing professional relationship. For the most part I was not a confidant of his, although I must say, in retrospect, in two or three months prior to the time that I became his successor, I got called in more and more on issues that were not within my realm of responsibility. I would be the only staff officer that would be out of step on things that the rest of the Army staff wanted to do. I particularly was at odds with the Deputy Chief of Staff for Logistics a number of times. But this is sort of historical and classical.

Once you were up there in the hot seat, as Chief, that put the burden on you to choose people for top jobs. How did you go about selecting generals for the three- and four-star jobs that came up?

This was not peculiar to me. For many years the Chief of Staff had annually sent out a list of two-star generals and he asked his three- and four-star generals to return to him the top five—I think I went to ten—the ten people from that list in the priority that they should be promoted to three stars. Well, this of course then would set up enormously different visibility as a three star than as a two star in most assignments. It was just a matter of personal assessments and in some cases personal chemistry and in some cases trying to avoid the impact of personal chemistry. You've got to look at capability, you've got to look at chemistry. Obviously you can't have a whole host of people around you with whom you have only a formal communication. You've got to have pretty generally an informal relationship so that you get information on things that don't necessarily come in official papers and this type of thing. You want somebody who will pop open the door and say, look, this has happened or this is about to happen and I'm doing this or do you want me to do anything or do you have some direction for me or this sort of thing.

But here again I go back to our earlier conversation where I said that one of the virtues of the school system was you're visible to an awful lot of people. You've formed a lot of associations. By the same token, this became advantageous in looking over the people that were available. And here's where I say you had to be taught to be careful of reverse chemistry. You have to avoid damaging somebody with whom you were not necessarily very familiar. I tried to be objective. I could afford to have no friends in that position. I tried to have two kinds of people in the staff immediately around me. One, I tried to have one person at least that I could call in and really bore, perhaps, for an hour, hour and a half, or two hours while I rambled around a lot of subjects, sort of thinking through out loud, watching for the reaction. Then I tried to have somebody, "personal staff," who tended to be, well, unkindly abrasive, but in reality, very direct.

I'm just a little bit curious. The people that were confidants or people that you could talk to about a variety of subjects were closer to you in rank, but people that you used for the sounding boards were much lower in rank.

We had an Army staff meeting once a week with all the principals of the staff and I encouraged frank and open discussion and exchange. I was briefed for Joint Chiefs of Staff meetings three times a week so that I had an association there. I was not an isolated person—nor insulated. I had lots of opportunities and I relied on my principal staff members in their areas of responsibility. This was not second guessing. This was sort of reaching out for perhaps other things. One of the things that disturbed me was the civil military relations in the hierarchy of the Pentagon. Well, this isn't the kind of thing that you tossed to a staff division to study. In the first place, there's too much exposure of it. In the second place, because it's an unusual subject, essentially a controversial subject, it's not the kind of thing you want a staff study done on away from you because you're not going to get a good product out of it. You're going to get a watered-down thing. You're the victim of the compromises that take place at a number of echelons below you so that you get, in effect, the least common denominator. Well, I wasn't trying to get the least common denominator in these sort of special issues.

And there's another aspect of it. I tried, unsuccessfully I suppose, for about three of the four years to instill a program of counseling by commanders at whatever echelon with their subordinates, a relatively informal exchange. I tend, I think, by and large to be a relatively informal person. This is just maybe one of my characteristics. Yet I sort of like a formal structure so you know where people stand and who's in charge. Basically, somebody's got to be in charge or nothing gets done.

Here's an example: All my life I felt that if the Army day ended at 6 o'clock it would be great. A lot of the social activities get to be pretty darned boring, talking about kids and that's about the only common denominator that you have in a lot of cases, but you never get into any discussion, very often, that is intellectually stimulating, so it's sort of a waste of time. I got a proposal that came right from the commander-in-chief over there and I fired it back and said, "I'll give you one reception and one dinner. The rest of the time I want to host a dinner at various places in the command for company level commanders, whether they be captains or lieutenants, where you can have one or two more senior officers. If it's a larger gathering, you can have the battalion commander, brigade commander. By and large I want this to be the people that are dealing with the troops, they're the people I want to talk to."

Now, what I'd do is, we'd have dinner and you have a drink or two before dinner, you'd have dinner, then I would talk—depending on the group and general mood and so on—very informally, just off the top of my head, all about what the Army was doing and how I thought it was going. We were heavily involved in Vietnam and everybody had a real concern. I'd bring them up to date on what I'd seen the last time I was out there. I went out there about two times a year. Then I would invite questions. Well, I didn't get any questions. So I'd say, "Look, I know you guys sit around over here at night and say, why did they do this and why do they do that? I'm they. What are you really worried about? Let's see what we can address. Ladies, you're invited to ask questions too." Well, you didn't get many from the ladies, one or two. "Why do you have only good beef in the commissary and not choice?" Found out why, everybody had changed the rules three or four years before because beef was out of price for the E-5s so they got beef that the E-5s could buy and everybody bought the E-5's beef. Everybody had to buy that—you didn't have any choice.

As I say, I was not successful in instilling in people really the essentials of counseling. I think many of the problems that bubbled up could have been forestalled by some explanation. But it takes time and it takes effort and it's risky lots of times because you're exposing yourself to criticism and yet in certain instances we went overboard and the other way. Several years later out at Ft. Carson in 1969 or '70, where General Rogers was then commanding the 5th division, he went over to the enlisted club under a purple light and sat in a chair where everybody else around him could see. It was dark. He was under sort of a blinding light. This was the mode of the moment, this flashing type of thing, and he responded to questions.

He purposely didn't want to know who was asking the questions?

He didn't want to know who was asking the questions. It was designed so that there wouldn't be any inhibitions about asking the question, which is another way of looking at it.

I suspect that there were times when you were personally frustrated by the Vietnam War. Is that true?

Well, it's understated.

Understated. Can you shed some light on that or how you felt about that?

My frustration, I guess, was that I believed then and I believe now that we could have solved that problem successfully. Whether we could have solved it within the agreements to which we were a party, I'm not that certain. By that I mean, we had signed the '54 Geneva Accord. We had signed the 1962 Geneva Accord in which we had pledged not to violate the neutrality of Laos and Cambodia. Yet other parties signed the accords who violated them with impunity and it seems to me that there was no real bar to a diplomatic effort to bring this out and then we take appropriate action and do what we had to do to bring that to a conclusion. We could have done it without an invasion, I believe, of North Vietnam. Invasion of North Vietnam, the overthrow of the government up there, the seizure of the governmental complex would have made it terribly difficult for them to continue directing the war and it would have made it impossible for them to provide organized forces for injection into South Vietnam, which they did. Second, barring an invasion of North Vietnam which would have been a terribly sensitive political problem and there would have been very legitimate arguments that this could trigger a Soviet or Chinese reaction.

However, I think again that when you undertake a task you have to be prepared to assume the risks associated with the task. We were never prepared to assume the risks. We tried to keep it riskless as far as outside was concerned, yet we tended to ignore the enormous costs, and I'm talking here in human beings, that paid for having it riskless in effect; in other words, not getting a reaction from the Russians, not getting a reaction from the Chinese. Now it's unlikely that the Russians had very much capability to do anything in the Far East. The danger then would have been, what sort of reaction might have been in Europe and our ability to cope with that reaction. I held then, and I hold now, the belief that nothing succeeds like success. And had we been successful the difficulties with our allies who were trying to teach us how to suck eggs would have been minimal. Sure there would have been a surge for a short while, then that would have passed.

Did you aggressively seek avenues to achieve some of these goals?

Yeah.

And is this what you encountered?

The Joint Chiefs as a body encountered it within the administration and I don't know where all the obstacles were because the Joint Chiefs of Staff were represented by the Chairman and just what his interchanges were I don't know. General Westmorland possessed a number of insights of which I was not aware. And of course an admission that is somewhat damaging to me personally—I acquired the feeling, the sense, that I was an observer, I was not a participant—particularly in my role as a member of the Joint Chiefs of Staff. Now I never did sense that as far as my role as Chief of Staff of the Army was concerned, because we worked hard at innovation and improvement of equipment and the quality of support and the improvement of the quality of the force. Despite our efforts to improve the quality of the force, there wasn't any question that as we enlarged, everybody that we got—and this is something that is overlooked, simply not thought of—that everybody we got, with a minuscule number of exceptions, came in at zero experience, both officers and enlisted men. We enlarged the officer strength from the neighborhood of 106,000 in 1964 to in excess of 166,000 in 1968. And every one of those people, with the exception of the doctors and lawyers and specialists of one kind of another, were zero-experience people as far as military was concerned.

How do you want to be remembered as Chief of Staff?

I think at the time that I retired I was asked this question. I tried to be the Chief of Staff for all of the Army, not for any special interest groups, and not to be representative of any special interest groups. I believe that I was successful in that. I don't think that anybody could point to me and say that I favored one group over another. I think that people would point to me and say that I didn't favor some groups enough. I hope that's not a contradiction in terms. It's not intended that way. And then the other thing I take enormous satisfaction in, I think that I walked out of the Pentagon my own man. I had not been captured. I had not been manipulated to do things that were not the right things. There were a couple of things that I'm not very happy about. I almost got captured once or twice, but I think that I avoided that. I gave it everything I had. When I started I said, you only do this once. I programmed myself for 14 hours a day, in at 7:30, five days a week and about a quarter to eight or eight on Saturdays. I made no effort to leave before 9:30. If I got my work cleaned up, I did.

Nine-thirty at night?

At night. When I got there, I had had my breakfast. I didn't go to the dining room. I did have lunch in the Pentagon. I had a mug of soup like this. When I got home at night, a plate of crackers and cheese. That was my dinner five days a week. I left on Saturday when I could get the material that was left on my baskets into two briefcases, one of which I carried in each hand when I left and I worked on them on Sunday. But I worked, I worked at it and I did nothing else but that. I bent every effort, no recreation. I don't play golf, I don't play tennis. I walked up and down the street with the dog after I got home at night for 15 or 20 minutes. That was basically it.

General William A. Knowlton

William Allen Knowlton was born in Weston, MA, on June 19, 1920. Commissioned a second lieutenant of cavalry from West Point in 1943, he saw action in the Battle of the Bulge, earning a Silver Star. Following the war he had extensive staff duty on the Army General Staff, at SHAPE headquarters in France and in Vietnam. From 1970 to 1974 Knowlton was the 49th Superintendent at West Point. He then served as chief of staff of the European Command and, promoted to four-star general, commanded Allied Land Forces in Southeastern Europe in 1976 and 1977. His final assignment was U.S. Representative to the Military Committee at NATO headquarters in Brussels. Knowlton died in Arlington, Virginia, on August 10, 2008.

10

"INTEGRITY IS AMONG THE MOST NECESSARY OF QUALITIES"

A conversation with General William A. Knowlton, USA
October 29, 1986

What do you consider to be the key ingredients a successful leader must possess?

Integrity is among the most necessary of qualities. Men have made general officer rank without integrity, but general officers are not anonymous and lack of integrity becomes far more devastating and far more visible in general officer rank. Those who do not demonstrate integrity normally do not rise to three- and four-star rank although they may fool people momentarily at the lower levels. If one is married, the leadership qualities and empathy demonstrated by one's wife are another vital component. Again, from the councils of the great, I know of cases where officers have demonstrated outstanding ability but their wives have not adjusted to the restraint required by responsibility. This quality sometimes does not become visible until a man becomes a general officer. The ability of a wife to handle community leadership and international social representation is another important facet despite the efforts of those involved with the feminist movement to denigrate it. Service in the Army remains a team effort. A thorough grounding in the military profession is essential, as is breadth of interest and ability. Specifically, assignment among three- and four-star generals is very often related to "musical chairs," and the timing cannot be controlled. Thus, a general may be assigned to a specific post that is not the one for which he is best suited, but because he was available for the posting when the assignment opened.

General, how do you handle difficult objectives?

The matter of handling difficult objectives or taking on difficult problems is handled differently by different people. In my case, the process involves two phases. In the first phase, unless time is terribly pressing, I dispose of the peripheral and lesser matters which may be up for consideration at the same time. This does not apply to the battlefield where quick decisions often have to be made. That clears the trash off the desk and enables me to concentrate on the most important issue without clutter. I also think instinctively, perhaps in my subconscious, about the principal problem while I am clearing away the trash.

In dealing with a difficult objective I try to break the problem into component parts within my own mind, and wrestle with the components. I believe strongly in getting advice, particularly on those subjects I do not know well. As Superintendent at West Point, I often called in cadets on some of the tougher issues of the turbulent early 1970s. While I did not always follow their lead, it helped me to know the perception of the young and those whom my decisions would affect. Quite often I took contrary positions, but only after insuring that I understood the view of the opposition completely, and after weighing it against my course of action.

I believe it important at the upper command levels that one understands the restraint required to keep from interfering in the detailed work of subordinate echelons. When dealing with an important issue on which my emotion runs high, I often dictate my views or actions and then put them away for several days. Later I may modify the language considerably or even not take the action, but it helps to get the honest opinions down as long as one can reassess later whether emotion will be helpful or not.

I believe in a combination of management by exception and the old

Armor philosophy of giving a subordinate a mission but not telling him how to do it.

Things which are not working, I look into in detail in the hopes that I can help correct problems. Those things which are working, I leave alone, beyond determining who is responsible for their working well. At the higher ranks, one must be able to deal with many issues in the same time frame—"keeping several glass balls in the air" to use the jargon. That means not getting so immersed in one issue that the others pass points of decision.

Admiral Arleigh Burke was known as a workaholic who really pushed his people, yet he would do things like leave a bottle of scotch on someone's desk as a way to say thank you for the extra effort. Do you reward your people for extra effort considerably above the norm?

The answer is "yes," but there is another part to the problem. In these days of the Freedom of Information Act, open files and multi-page copies, too many officers are afraid to categorize or prioritize those who work for them. Particularly in the upper ranks, I think it is important that leaders highlight carefully those who are doing an exceptional job clearly beyond what would be expected. This would involve outstanding efficiency report notices, special letters and often decorations. However, those actions gain credibility only if one takes actions also to identify those that have demonstrated the truth of the "Peter Principle." Officers who have reached their peak or who are not capable of carrying out their functions at the level demonstrated by others in the same position should be clearly indicated. My stepfather used to say "The Army is not a charity institution." In keeping with that, it is better for the individual and for the Army that those who lack competence be so indicated and moved to other duties. During my time in combat in Vietnam, I relieved two of my battalion commanders. This was not in response to pressures from higher headquarters or because of any statistical reason. Both had been outstanding

General, how do you find out what is going on in your command, particularly in the lower ranks?

My first commissioned service in the 1st Battalion, 40th Armored Regiment, under Lieutenant Colonel Ed McConnell was a revelation. With a few of the projects given me, I discovered that he knew more about the internal workings of his battalion than anyone suspected. It was at that time that he told me that a commander should always set up his own intelligence net. There are two tricks to this. The first is that the people in the net must not realize that they are in an intelligence net. The second is that information thus gained should not be used overtly for fear of "blowing" your source. But since the chain of command is notoriously slow and favorably colored in producing information, a commander must have an alternate means of knowing what is happening below. A key element to this is accessibility. I am fortunate to be the recipient of letters from a variety of officers, from lieutenant to colonel, who confide in me, talk out issues with me, and give me a feeling for the view from their level. The members of my immediate official family are just that. I speak very frankly in front of them as I think out problems, but with the understanding that the information remain within the office. In turn, I trust them to speak frankly to me when I ask their opinion on personalities or situations. This sometimes causes nervousness in intermediate level commanders. But one of the reasons I am so proud of the number of former soldiers under my wartime command who continue to write to me is my belief in the adage that you can fool those over you, but you cannot fool those under you. Obtaining information often means that one must ask questions.

I think the most important characteristic here is a genuine liking for people, and a liking for people that is apparent to others. This is hard to simulate, and I feel fortunate that I enjoy meeting people and knowing people. Initial calls or final calls in my office normally run overtime and normally are very informal. I try to keep notes on the officers in my command so that I can review them before conversations and demonstrate an awareness of what we talked about before. For example, I have a little book with notes on every General Officer in Turkey and his family. I review these notes before visiting a Turkish headquarters or area, and find that they are flattered when I remember past details and names. This, in turn, leads them to speak with me much more frankly than they would talk through the blanket of protocol. In summary, I would say that a high-level commander must make himself approachable and must make people feel relaxed when they are talking to him. The truth is much more likely to come out. Unfortunately, I must quote from an absolutely outstanding colonel who recently retired voluntarily in order to take an important civilian position with a university. He had serious concerns about the Army and about its future as revealed to him from experience as a brigade commander. We talked one day just before he left the Army, and we spent over an hour discussing the issues. At the end of this, he blurted out, "You are the only general to whom I have been able to talk on these matters. I either cannot get at the rest, or when I do, they walk away." If that is true, it is tragic. It also underlines the word accessibility.

Are you a commander or a manager, or is there a distinction between the two?

There is a distinction between a leader and a manager. A manager's results are measured in more quantifiable terms. A leader's results are often more important in non-quantifiable factors. What I am saying is that a good manager may also be a good leader, but a good leader sometimes has to violate the principles of management. It was about 20 years ago that the *Wall Street Journal* had the following quote: "It is a good thing when a General goes about the business side of his profession in a businesslike manner, but there is an element of the irrational in warfare that is not susceptible to business management." Some of the most destructive things to the totality of the Army experience have been justified as being good management. Do not interpret this as meaning that a leader must be wasteful of resources. Rather, a leader must understand and assess non-quantifiable factors as well as quantifiable. His interest in the two is in a different ratio from the view of a manager.

Every general likes to think of himself as both a leader and a manager. I am no exception, but I would weigh a little more heavily on the leader side. Here my wife has played a most important part. While I have been known as a troubleshooter on the military side, my wife has gained a reputation for the same thing on the distaff side. There are thousands of other things which could be said and are said in many leadership lectures. But the most important is to remember that military service is just that: service. The higher one goes the more essential it is to remember that one important capacity of the higher graded officer is "the ability to suffer fools gladly." This is tough to inculcate, but an essential piece of getting on with what must be done. Lord Chesterfield once wrote in a letter to his son something like this: "'As you grow older, my son, you will be amazed with how little wisdom the world is governed." It is the job of the military man in the higher echelon to see that those "of little wisdom" are given the wisdom and understanding to accomplish their job in a better fashion. But this requires loyalty, not only to the individual, but also loyalty to the United States as a larger entity and to its goals. Service requires dedication. Dedication requires forbearance and understanding on the part of family and friends. But the end is worth it in terms of satisfaction.

General Knowlton, what qualities do you look for in a subordinate when selecting him or her for a particular assignment?

I have not been one who takes an entourage with him to a new post. Basically, I accept what I find on the ground until a reason for change manifests itself. At the same time, trained by General Gruenther, I require high standards of those who serve for me. The qualities that I look for in a subordinate depend to some extent upon the position for which I am looking. Let me give an example. Among the many problems which awaited me upon my arrival at West Point was a post staff which was not functioning well. As a former professional SGS (Secretary of the General Staff), I realized that my SGS was not adequate, and that a good man in that key post could change the entire staff. Each time the SGS got in difficulty, he added someone to his staff. I called in Lieutenant Colonel Thomas E. Fitzpatrick, who had served with me in Military Assistance Command, Vietnam (MACV) and again when I was SGS. He had certain qualities which I was looking for: toughness, ability to work quickly and simultaneously on several items, understanding of people and organizations, staff ability, exceptionally good judgment, an ability to work without detailed direction, and a knack with people that ran counter to his toughness.

In MACV, when General Westmoreland was put in charge of the civilian agencies in 1967, Fitz was able to require high standards from the civilians, far higher than had been required before, and yet made them all admire him. In one year he straightened out the West Point staff and cut down the size of the office he inherited, with a great gain in efficiency. I then shifted him to a cadet regiment with which we were having difficulties and where I sensed the source of problems which erupted last year. The two previous regimental commanders had not been able to get a grip on them. In one year, Fitz was able to turn the regiment most of the way around. He then moved to become the Deputy Commandant. In both jobs, his knack for handling the young and of being able to communicate with the young was the key to his success. That had not been required in the SGS job, but it was a requirement for the job with the cadets.

I cite those to demonstrate that the qualities required in different jobs are different. I guess basically I look for someone who will be completely honest in his statements to me; he is not useful if he does not do that. He should be able to take broad policy guidance and apply it in detailed specifics without the need for more. His judgment must be good, for bad judgment can be disastrous. He should be able to relate to people, have an appeal to those who are younger, and not be afraid of handling questions. At the same time, he must understand the limits of popularity and be prepared to be tough when the occasion demands. He must be a professional in his detailed knowledge of his profession and in his ability to apply it. At the same time, he should not be single-mindedly devoted to the profession to the exclusion of family and hobby. Not only are they relaxing, but a man without family and hobby cannot understand the problems of those who do have them. Loyalty is a prime requisite, and does not run counter to the need for honesty.

It is my habit to speak frankly in front of my immediate staff and to think out problems out loud. In those circumstances, breaches of confidence can be devastating, so loyalty and an ability to keep quiet when necessary are most important. Those are only a few of the things which make up my instinctive judgment on subordinates when I have the option of requesting them. I reiterate that the qualities vary according to the position. There is one officer with whom I have served on two occasions who is a good example of this. In Vietnam, he took over a dispirited battalion of indifferent success. Demonstrating great personal bravery and calmness in distress, he turned the battalion around. Yet, in another situation, he found a staff position in a higher headquarters difficult to do well. An outstanding individual with a lovely family, the calm and relaxed nature which he demonstrated on the battlefield ran counter to the nervous energy required by a totally different position. So, there is no overall response to your question, but a character assessment based on job requirement.

General Frederick J. Kroesen

Frederick James Kroesen was born in Phillipsburg, New Jersey, on February 2, 1923. He dropped out of Rutgers University in 1942 to enlist in the Army and received a commission as second lieutenant in 1944. An airborne officer, Kroesen commanded every size of unit from platoon to army, including the American Division in Vietnam, the 82nd Airborne and VII Corps. He served as a four-star general from 1976 to 1983, when he retired. General Kroesen is active in the Association of the U.S. Army and resides in Falls Church, Virginia.

11

"TELL PEOPLE WHAT HAS TO BE DONE AND NOT HOW TO DO IT"

A conversation with General Frederick J. Kroesen, USA
October 21, 1987

How do you describe your own style? What is the Kroesen leadership style?

Leadership is a complex subject. Essentially I have trust and confidence in my subordinates. I believe somewhere, sometime in the Army they have impressed people who have promoted them to the positions they now hold. That goes for a man who is a buck-sergeant, that goes for a man who is a full colonel, that goes for a man who is a major general. I may have never seen these people before but because they have qualified to be in the position they're in, I accept them at face value and expect them to carry out the responsibilities of the job they've been assigned. That starts us on an even keel. I feel they're qualified; they don't have to prove themselves to me. All they have to do is do their job well. I want them to know that I accept the fact that they're qualified to do it. Now over time, if he doesn't do it, I am quick to remove him. Not necessarily based on the fact that he made a mistake if he doesn't make the same mistake twice.

Now, that brings me to the next real cornerstone of my philosophy. I believe in mission-type orders. I believe in telling people what has to be done and not how to do it. When I give a subordinate commander or a staff officer a job to do I do not stand over their shoulder and tell them how to do it. I allow them to employ whatever innovations they want, make the mistakes that they're going to make, and learn from it. I'm happy to critique them and tell them what they did wrong or tell them how I think they could have done it better, kind of a part of the education I think is a commander's responsibility, particularly in peacetime. So I believe in the decentralized kind of directives. Tell people what has to be done and not how to do it.

Is the Army promoting its generals too rapidly and retiring them too early?

As a matter of fact I don't disagree with the Air Force policy of retiring people even earlier than the Army does but I would not want to have been retired any earlier than I was. I think I was still making a reasonable contribution and that my experience was worthwhile to the Army. So I believe in the Army's system of selecting, say, five people who stay on beyond 35 years, who can be on even longer. That's what General Lemnitzer and General Bernie Rogers did, both of whom were kept on beyond the normal retirement date of thirty-five years. But if you don't have an early retirement requirement, you stagnate the promotion system for the younger officers and you end up with an officer corps that is too old for the job it has to perform when war comes.

You've been studying the history of the Civil War. The officer corps, particularly the general officer corps of the United States Army, was much too old when the Civil War started. And the younger officers who went off to the Confederacy and became generals in the confederacy at a younger age, certainly carried the South in that war for the first two years or so before the Union Army was able to weed out and promote properly and got people in charge who could fight that Army most effectively. If you look at those armies today that do not have an up-or-out policy, and the German Army is a prime example, all generals have to stay until they're sixty years old. That was a fine policy perhaps when they first started but by 1985 what they had was a whole list of generals waiting to become sixty and a whole bunch of majors, lieutenant colonels, captains, who can't be promoted because there are no vacancies at the top. And you end up in Germany today with infantry battalion commanders who are 45, 46 years old. That's the age we want brigadier generals to be in command of forces. The whole Army gets too old for the job it would have if a wartime crisis occurred. We can't afford in our Army 45-year-old battalion commanders of infantry, armor. They are physically and even mentally not attuned to that level of conduct. So if this Army has to give up because of congressional pressure or whatever, if we have to give up our up-or-out policy, you'll see stagnation of promotions and a stagnation of capability of the officer corps.

But isn't the only risk, General, that the quality has to remain consistently high so that you always have two or three deep the candidates who can step into those one- and two-star jobs when the vacancies open up?

We always hear of this requirement by the personnel managers that we've gotta train so many battalion commanders so that when the Army is mobilized, and we go into full expansion in wartime, that we'll have enough experienced people to take care of that expansion. They think in terms of World War II when we had 75 divisions, or whatever it was. Yes, it's ridiculous. In the first place we can't equip 75 divisions today, we couldn't put that many in the field. In second place, we don't use people who have already proven themselves in one job and send them back to it. You can't find me any battalion commander today who has formerly been a battalion commander and is now doing the job for a second time. You can find them—there happen to be three of them that we made exceptions for. The Ranger battalion commanders, one of the requisites is that they have formerly successfully commanded a battalion. I was one who helped put that policy into effect. When we select ranger battalion commanders they need to be people who have proved themselves already. But aside from that, and if you look at the Vietnam War, we didn't train people as battalion commanders in the United States and then send them over to be battalion commanders in Vietnam. Everybody in Vietnam was doing the job for the first time, including battalion commanders, brigade commanders. You didn't have the experience of being a battalion commander available to command battalions in Vietnam. The idea was, you gotta pick this young lieutenant colonel who has never been a battalion commander and is given his opportunity. Well, that's the way the personnel management system mainly works. So I rate two or three deep every year, promote about 50 people to be brigadier general. The board that promoted them lists 200 colonels who are fully qualified, could be promoted to general officer, but only 50 of them are. The other 150, some of whom will make it next year, some of whom will never make it, are identified as qualified, capable of putting on stars today if you had a reason to do it.

General, looking back when you were a junior officer and you had a certain perception of four-star generals, did that perception change years later when you were one?

I never had a perception of four-star generals when I was a young officer. I guess I was a great admirer of those officers who fought World War II successfully. I thought very highly of General Eisenhower, General Bradley, General MacArthur, General Patton, because of the things they accomplished. As time went on and I was able to read some of their biographies or the history of that period, I found that some of them had clay feet, some of them had flaws and foibles. So I came to the realization that they were ordinary—not ordinary, maybe—but I never had any hero worship for such people. I guess when I became a general it never occurred to me that anybody would have that kind of feeling about me. I didn't change my opinion of what generals were. I believe generals are people who are sincere students of their profession and who then either have the inspiration and artistry to excel in times of crisis or they don't have when the time comes. And those who do have are mainly those that you know, that you hear about, that are successful. Those who do not have that artistry, are not able to combine that art and science in the pursuit of their profession, are the ones you never hear about. They've gone by the wayside, they are relieved of their commands, they are given backwater positions that are important but not the limelight of leadership.

During your years on active duty, particularly the later years, who among your contemporaries impressed you the most in terms of their leadership ability?

Well, I don't think I want to pick any of them out. I'll tell you that I have very high respect for General Jack Vessey, I have very high respect for a number of others for the particular responsibilities that they've held. For example, General Bernie Rogers is almost an ideal selection to be the Supreme Allied Commander in Europe because he has a background in political and international affairs that allows him to be effective in an international field. Other people like General Don Keith have an affinity for dealing with the industrial complex of this nation. I respect him very highly for his ability to command the Army Materiel Command. I think all the services operate with sort of an executive committee of four-star people, all of whom have special responsibilities and are usually a good selection for the job they have to do. I think General Kerwin was an ideal Vice Chief of Staff. He's a detail man, a guy who could sit behind a desk and coordinate the activities of the Army staff and get that kind of a job done. I don't really know that much of General Kerwin as a field commander, but I know I respect him for the function he was performing and the way he did it.

The Army has about seven four-star generals who run the Army. It has others who serve as commander of Readiness Command, now the new Special Operations Command. It's got some who are off on joint duties, but there are seven of them, maybe eight, who run the Army commands and the Army staff. They are an executive body, each of them doing their own job, none of whom feels subordinate to or inferior to the others because of the position he's in, including the Chief of Staff. I, for example, as the Commander in Chief in Europe never felt subordinate to the Chief of Staff of the Army. I was known as one of the major commands of the Army in the world. My job called for a four-star general just like his job called for a four-star general. The Vice Chief of Staff is a four-star general. I didn't feel I was being demoted when I left the four-star job as the Vice Chief and went to become CINCUSAREUR. I wouldn't have felt demoted if I went from the Chief of Staff job to Europe.

So, all of those four-stars I have known had a particular function to perform and my consideration of them was how well I thought they were doing that job. Most four-stars only get one job. I had three of them. Some have two if they go from a four-star job to the Vice Chief or the Chief. I was lucky enough to have three different four-star jobs. There aren't many who do and most only have one and they put in their tour of duty in that one job and then it's time for retirement. I can name you the ones who impressed me the most in World War II and other people like General Abrams and General Westmoreland. For the particular job he had to do I don't think General Abrams had any superiors, certainly, maybe not even any peers. He was an outstanding leader and one of my favorite people. I wouldn't count him as a contemporary. I worked for him. He made me a division commander in Vietnam.

How often in the cycle do leaders of the Abrams' ilk come along in the Army? Or are they there all along just waiting for the right opportunity or, perhaps more importantly, waiting to be discovered?

I'm inclined to think that they are there most of the time, waiting to be discovered, but they are not always discovered. They are victims of the age in which they live. If you think of the generals who were in the United States Army from 1919 to 1939, that's a twenty-year period, there were some very good men who came and went in the general officer corps during that twenty years. You don't know how many of them were Pattons or Eisenhowers or John J. Pershings. They just didn't happen to be generals in the right twenty-year period, or the right five- or ten-year period. We had a war, '17 and '18, and we had a war from 1941 to 1945. If you weren't the right age group—for example, the class of 1915 turned out a tremendous number of generals who became famous in World War II.

And why? Because they happened to be 1915 and one or two of them got promoted early and they then brought their contemporaries along. When Eisenhower had to pick somebody to be a three- or four-star general he knew the class of 1915 best. And so they had a leg up on selection. Doesn't mean they got selected because they were the class of 1915 above other reasons, and it doesn't mean that someone better qualified couldn't have been selected in war in many cases. But the fact is that if you have two equals and the only difference is you know one guy and you don't know the other, you select the guy you know because you have experience and know you can count on him.

You had a reputation in the Army as a somewhat fearless leader. How did you feel about your own mortality?

Well I have certain fatalism about my own mortality because of the profession I've been in. You see, I almost lost my life as a second lieutenant on more than one occasion. I was the only officer in my rifle company who went overseas with that company and was still with it when the war ended, and that was only seven or eight months later. I was standing next to four of the officers in my company when they were shot and wounded bad enough to be evacuated and I wasn't touched—standing elbow-to-elbow with them. One occasion, two of us were walking down the street. He was hit and I was not. So that gave me, right at the early part of my career, an 'if it's going to happen, it's going to happen' realization. I've never brooded about it and it's never been of great concern to me when I've had to go into combat or do anything else stressful. I've been a paratrooper, and people get killed parachuting, but it has not been one of my concerns when I jump out of an airplane. Whether it's a Middle Eastern fatalism or whatever, it's sort of incidental what brought it on, but that's the way I feel about it. I feel the same way about the terrorist attack. He had a good shot at me and he missed. I don't know who I have to thank for that but I'm willing to believe it's been some sort of divine protection of my well-being for the 40 years I've been in the Army. I've been wounded three times, never badly, just sort of enough to get a Purple Heart.

General LeMay was known for sitting his new aides down their first few days on the job and saying look, the one lesson that I want you to learn in this job is to keep your eyes open and see what I do and pick up my strong leadership skills. Did you operate like that with your aides?

I don't believe in passing on any leadership style because I believe every person needs to have his own. If it is not a natural style for you it's going to be perceived as a manufactured one. All great leaders have their own style of leadership. The difference between Patton and Bradley was so marked that if either one of them tried to pattern themselves after the other they would both have been failures. There was a wide-ranging scope of behavior and leadership characteristics that can be chosen from or which you naturally employ.

What about yourself?

At the same time I have said this in different words than you use. The legacy that the officer leaves is the people who have been trying to take that guy's place when he departs. It's important to me that my legacy to the Army is the people that I helped to train to do the jobs that I did in a succession of positions who will take my place in the future. I hope they do a better job than I did and I hope that one of the reasons is because I helped to create the foundation upon which they could take off to do a better job. I have said in all of my leadership conferences when I talked to battalion commanders who were newly assigned to Europe—I used to do this once a month or so depending on how many came in—we ran a school for them and I always attended the school and gave my little talk. I said to them, we don't know whether we're going to go to war. Your job is to be ready to go to war tomorrow, if it happens. But more importantly, because none of us foresees a war coming in the next two years—which will be your tour of duty—more importantly, you leave behind the foundation for the next guy who will do a better job than you did because of the battalion leadership that you gave to that battalion while you were here. I believe in that. I think that there is that attitude requirement of the officer corps and of the Army to assure that the next generation will be better than they were.

General Curtis E. LeMay

A legendary proponent of strategic airpower in World War II and for decades thereafter during the Cold War, Curtis Emerson LeMay was born on November 15, 1906, in Columbus, Ohio. He earned his wings in 1928 a year after graduating from Ohio State University and spent the next twelve years as a lieutenant. A captain in January 1940, LeMay became a brigadier general in October 1943. As head of Strategic Air Forces against Japan he earned promotion to major general and, by October 1951, had catapulted to four-star general—one of America's youngest at age 44. General LeMay spent the next fifteen years as Commander-In-Chief of SAC, Vice Chief and Chief of Staff of the Air force. He retired in Southern California and died on October 2, 1990.

12

"NEVER MAKE A DECISION UNTIL YOU HAVE TO"

A Conversation with General Curtis E. LeMay, USAF
August 16, 1976

Leadership is a discipline that has been discussed and debated since the dawn of history. Are there certain rules that apply to leadership development?

There aren't any rules of leadership that I know of, except maybe the golden rule. I don't think this is something that you can be taught. Say do this, do this, do this. Some people would do that and they wouldn't get anything done. Other people do it and they do fine. I don't think you can tell anybody how you can go about being a leader. That might be my difference in ideas or something that they might work on. First, if you're going to lead somebody or do a chore, it might be helpful if you knew more about it than they did. There's too damn many experts now running around that know absolutely nothing about their subject, but qualify themselves as experts and trying to get something done. And they'll find a bunch of sheep that will follow them around, too, if they talk with them enough. But they don't get much accomplished.

General, wasn't there a certain amount of fear in your leadership style? It's been said that you put the fear of God into the people that worked for you.

No, no, no. This is a bunch of crap. I never scared anybody in my life. I don't think anybody is afraid of me. They're probably afraid of themselves. Everybody knew what my standards were and they were high and if they were afraid of me, they were afraid they weren't meeting my standards, that they weren't doing the job.

How much tolerance did you have for people who weren't meeting your standards? I've been told you had little or none.

Well, it depends upon how much time you have. If you're out in a war with people getting killed, that you're responsible for, you don't have a hell of a lot of tolerance for some stulpnagle that's not doing his job. You get somebody in there that can do it, and quick.

How does that roll over in peacetime? You come into a new organization, a new command. There are people there who are obviously not up to standard. What do you do about that?

Fire 'em. Get somebody that can. Talking about in the military, now. Let me ask you some questions. Who was the Undersecretary of State at the time of Pearl Harbor? Who was the Chairman of the Military Affairs Committee at that time?

I don't think I can answer those.

Who was the naval commander out there at that time?

At Pearl Harbor?

Yeah.

Well, you had Admiral Kimmel and I know General Short was the other one.

Right, General Short, you remember them, but you don't remember these other guys that were probably more responsible for that disaster than Kimmel and Short. But you remember the military guys that got hung for it. That's what I'm talking about. If you're in the military, you'd better be at war all the time. As a matter of fact, we are at war all the time now with Russia, whether you like it or not. Most people are too stupid to even think much about it. But they keep telling us all the time that they're going to bury us, that their goal is world revolution, and so forth and so on. They may not be shooting at us all the time, but you're at war nevertheless. All you good patriotic citizens demand instant reaction of your military; you expect us to go out and to fight, maybe not even win, like Vietnam. Just go fight, get yourself killed out there, while we stumble around back here and see what we're going to do. You've got a different situation in the military than you have in civilian life. You can be careless and slipshod in civilian life and you don't cost a half a dozen men their lives. At the most you just cost them a little money or something. You can do that in civilian life, but you can't do it in the military. You're better off getting used to it

While we're on the subject of civilians, General, how do you feel about the leadership in the Pentagon during the years you were there?

I don't remember being helped particularly by any of the secretaries or those people in those bureaucratic jobs.

Because they didn't want to help or they didn't understand?

They didn't know how to help to start with. By the time I got in there in the service secretary's office, I didn't have any authority anyway. It was all run out of DOD.

When you were picking people, General, during the eight years that you were there in the Pentagon—the last eight years of your career—what sort of things did you look for in your senior commanders? People that were like you?

No. I looked for somebody who first knew his way out to the can. He knew something about what he was supposed to do. So, I want someone to know something about what I want him to do, knowledgeable on the subject. I want him to have a sense of responsibility.

How do you define responsibility?

I look at what he's done in the past and how he's doing his present job. So as a rater you can look at somebody and you can tell, well, does he give a damn or not. When things hit the fan over there, he's going to be on deck to handle it. Doesn't take any great amount of brains to size that situation up, does it? Well, that's leadership.

Could I be just as effective a leader as a colonel as someone could be as a lieutenant, or somebody could be as a four-star general? Does it change any?

I don't think so. It changes as it goes up. If you've got two men that are digging a hole to plant a tree out in your yard, and you want to get them to do the job in the most efficient manner possible, that's one set of conditions. If you've got five hundred men out there reseeding a hillside that's been burned over and you want to get that done in an efficient and orderly manner, that's a different problem in leadership entirely. You might do the first one and do it right and fall flat on your face in the second one. Or vice versa, for that matter. All situations are different.

When P.K. Carlton was working for you, did you take it upon yourself to teach him some of your leadership styles or did he evolve as a leader on his own?

Well, I gave him some training, but he did it on his own. Early in the war I met an old man over in England, Lord Trenchard. He had commanded the British expeditionary air corps in Europe in World War I. He was responsible for getting the first independent air force in the world, the Royal Air Force. I guess he must have been pretty close to 80 then, but he was still a fine old man. He put on a uniform and spent all the time he could where he could walk—felt like doing it— got around to the bases, our air bases, American air bases and British air bases to see what the troops were doing just to talk to them a little bit. Tell them a couple stories. After a couple talks with him, it dawned on me that Portal, Tedder and the guys who were really running the RAF had been his aide at some time or other. So I remarked on this coincidence, how it had happened. He said, "Well, it was no coincidence. Nobody wants to be a dog robber for an air marshal very long. What I did was pick out some of the smartest guys I could find, keep them with me for a short period of time, and let them see what problems I had to deal with and how I went about solving them. Being a smart guy to start with, he learned something. And he retained it and went on and did well when he got more important jobs and so forth." I thought that was a pretty good idea. So I tried to do that, because I didn't have the freedom of getting people and determining their assignments as Trenchard had. To some extent I did pretty well. A couple of my guys became four-star generals, a couple, three others, major generals. One exceptionally good one was killed in action in Vietnam or he'd have been up there too. I did pretty well. Of course, I picked out some smart guys to start with. I tried to give them some training—the kind of problems that you get and how you go about solving them. Getting an answer, hopefully the right one.

How long did it take you to size somebody up to know if they're the right person?

Thirty seconds to two years.

That's pretty broad. You have a reputation for not much tolerance, General. If someone had potential—you felt they had potential—but early on they weren't up to speed, how much leeway would you give them to come up to speed?

It depends on the situation. Everybody is entitled to make a mistake once in a while. All the people that don't make an occasional mistake are people that don't do anything. That's the crime. When you've got a job to do and something should be done and you don't do it, that, in my book, gets you fired right now. You do something and make a mistake, okay, I might go with you if it's not too horrible a mistake. Somebody quoted me as saying, and I guess I probably said it all right, that I have trouble distinguishing between the inept and the unlucky. I didn't want either one of them around.

Do you feel that there's enough people to pick from that you could afford to be that choosy?

Normally you have to take what you're issued in the military establishment. In these times it's a little more trouble to get rid of them, but you can get rid of them. In wartime you get rid of them awfully fast. No matter how bad things get, you can always find somebody, if you dig enough.

You mentioned the golden rule. Some of these traditional concepts that you find in religion—you know, turn the other cheek and all those others—do they have a place in leadership?

Turning the other cheek is not the golden rule. Do unto others as you would have them do to you. I think this applies to most things in life and your relation with other people. You ought to be fair with them. Now you can set high goals if you have to, if you've got a job to do and you've got to get it done and it means hard work on the part of everybody, you can set the goals. This is what we've got to do. If they don't make them, get somebody who will. But you've got to give them a fair chance at it. You've got to let them know what's expected of them. What has to be done.

And how do you do that?

By telling them! By telling them! In other words, how do you want to be treated? If you've got a boss and you want to work for him, you'd like to know what he's trying to get done and what he expects you to do about it, and you don't expect him to fire you when he hasn't told you what to do and you haven't got it done. In other words, you expect fair treatment from your boss.

That seems simple. Why do so few people do it?

Well, I don't know, it seems simple enough to me. But a lot of people are not interested in giving the people under them fair treatment.

What about credit for a job well done?

If they do a good job, they'd be very glad to take the credit for it. That's not being fair either, is it?

No, sir.

A military organization is a team. Usually the commander gets credit if you do a good job. He's the top man. Say, okay, Joe, you did fine. When that happened to me, that meant that we, the outfit, did fine and I let them know about it. In presentations and things of that sort, you talk about my command, I did this with my command and so forth and so on, that's MacArthur's style. I don't go for that. My command was we! We as a team did it. It is a team effort. Everybody has to do his job and exactly at the right time and the right moment and everything matches together and gets it done. You get the job done then. Somebody doesn't do their part and the whole damn thing will probably collapse on you, particularly in a military operation. So, you operate as a team and you ought to pass out the medals as a team or you ought to kick the butts as a team, those who need it.

In order for the team concept to work you've got to know what everybody is doing at all these various levels. How do you communicate that? How do you, as the Chief of Staff of the Air Force, keep in touch with all these people down here at the bottom of the list?

You devise various methods of doing that, but you don't have to look at everybody every day all down at the bottom of the chain of command. You get some guys up here that you know will do the job. And say, okay, Joe, you do this. Go get it done. If you get the proper people, you can forget about it, knowing it will be done. In the military you're trusting your life on knowing that he will go and do that.

I guess what I'm saying, General, that's at the top of the pyramid.

You had eight lieutenant generals working on the E-ring and then you had all those major generals and brigadier generals and colonels—people that were making things happen for the Air Force. How did you personally know what was going on?

I usually laid out the rules and got the methods how we were going to do things. Sometimes down to the bottom detail. Okay, this is the way we do it. Take flying airplanes, for instance. We start flying this model airplane around and we find that we have some accidents. Now we kill some people along the road. Well, what happened? Well, this happened here and this happened here, this happened here. With this chain of dead people you've got along there, you come up with a method of operating that airplane. If you do it this way, you're liable to get killed. So you'd better do it this way. So, you have this method of operating. Say, okay, we're going to use this method of operating right up to the top. Then you assign somebody to carry it out. Or if there isn't any method and you haven't got time to find one out yourself, you say, okay, Joe, you do it. I don't care how you get it done. This is what I want accomplished, get out of here. If you've picked your right man, you can cross that worry off for the day, because you know he will go and get it done. If he can't get it done, he's going to be back beating on your door telling you about it.

Did you personally feel more comfortable picking people with a SAC background rather than those who were in missiles, for example? Did you tend to give more preference to those people?

No, not necessarily.

But you were more comfortable with SAC people.

I was more likely to know the people that have been in SAC, personally, I mean, rather than just someone on an efficiency report.

Can you pick somebody by looking through their file without looking at them, talking to them?

It's very difficult. One of the most difficult things that I had to do while I was Chief of Staff was pick the people for appointments for general officer grades. The opposition was pretty tough. I'd find people that I'd served with and knew, knew intimately, and competition was somebody that you didn't know personally. How am I going to evaluate them? The way I used to do it was, I'd have a promotion board. Say, okay, it looks like we'll probably be able to promote a hundred people to brigadier general this year. Convene the promotion board, divide the people in two: those that shouldn't be considered, those that should be considered. Then I get another promotion board and say, okay, you take this list of people that the other board said should be considered and you pick me out two hundred that should be considered. We're only going to make a hundred, you pick me out two hundred. I'd take that two hundred and I'd personally go over their efficiency reports. I had a big bunch of these market baskets full of records by my desk all the time and when I had time, I'd look at them. I'd look at them at night, burn the midnight oil. I'd finally pick out a hundred guys.

Using what criteria, General?

I'd do it myself.

I mean, what do you look for?

I'm looking for everything I can get my hands on. Their efficiency reports, their personnel records, letters of commendation, the jobs that they've held, their whole life. Then I'd have the Vice Chief of Staff do the same thing. Then I had the Chief of Personnel do the same thing—the three of us. And when all got through, I got them in the office. Let's see your list. Well, we'd look at the list and if some guy was on all three of our lists, he was in. Put him over on the list here. Then we picked the guys who were on two lists. I'd talk to the man who didn't have him on his list. Why didn't you have him on your list? Sometimes I'd find out something that wasn't on the efficiency report or I don't know why I didn't have him on the list, I just didn't like him. Then we'd talk that over and go through the whole business that way.

Then I would make up the list of a hundred, using that decision.

When you were Chief of Staff did you have an open door? Did you encourage people to come in and tell you problems? Isn't it very well known that you didn't like to hear problems, you wanted to hear solutions?

Well, you have to have a problem before you have a solution. You don't run around with a handful of solutions finding a problem for them.

Well, of course not, but did you want to have just the problem or did you want to come and say this is the problem and this is what I recommend we do?

You've got a problem first. If you've got some answers, okay, fine. Usually the first answers are not much good. You've got to think things over. A good rule to follow is never make a decision until you have to. But don't make it one second later than you have to. Have the decision on time, but the longer you delay the more information you can gather, the more input you can pour into the solution, the more likely you are to have the right answer.

So how do you know when you have enough information?

How do you know anything?

How did you know, General?

Well, how are you going to know when you're finished with me today?

I don't know. I'm going to have a sense that we're done.

You know something of what you're trying to do and you think you've got about all you're going to get out of me and you'll think you're finished. Use this as a matter of opinion and judgment. There's no hard and fast rules in this business. You may do things different tomorrow than you did today.

Sir, you are still very active with Air Force matters. Are the officers you encounter any different from those who you knew and worked with years ago?

Yeah. I think they are.

In what respect, General?

I think most of them seem to be a little better educated than we were. They've got a better military educational system. Hell, I come from a generation that never got a military education. I had a get-rich-quick course of three months in the Tactical School. I never got to the National War College, the Air War College, or the Staff College, Company Officer's School, or any of these specialized courses they have now.

But do those make people better leaders?

Why certainly. You get more training on how to solve problems. You get to read more military history. You get to practice some of the things that you have to do in these advanced jobs before you get there.

So, in other words, General LeMay, if people possess certain basic characteristics, they can be taught to be effective leaders?

You teach them how to solve problems, how to think logically.

But leadership is a lot more than that.

Yes and no. You have to make up your mind what you want to get done. If you don't know what you want to do, you can't lead. So, you've got to be able to make up your mind what you want to do. Then the next thing is getting people to do it for you.

There must be some sort of intangible there that I'm not grasping. You have any two men who are alike in training and education and in experience—what makes one man a charismatic, effective leader and the other not? What is the difference?

I've been trying to tell you that this is hard to define. There are no hard and fast rules. I'll pick out some people who were supposed to have been good leaders. Let's take MacArthur. I wouldn't follow that man out to the outhouse because I don't like him. I don't like his methods. Now there are other people on the other side of the picture. Patton and MacArthur are about the same. They're both actors.

What about somebody like Ridgway?

Ridgway, he's on the other side of the fence. The difference, then, is characteristics. I'd follow Ridgway. He's about like I am. We think the same. Act the same. I would expect him to be there when the going got tough. And he would be there. MacArthur would probably be down getting his picture taken some place. So there are all kinds of people. Mainly the troops want to know that the boss knows what he's doing. He's not going to get us into any impossible situation unless it's necessary. He knows what he's doing. And we know what he wants, what he expects out of us. And we know that if we give it to him, we're going to get promoted, he's going to take care of us. And if we don't do it, we're going to get fired.

Do the men have the capability of identifying a general who might be a sham? Do they know that?

Sure they know it. If he is a sham, he gets them in a jam. And you learn pretty quick when you get your ass shot off by some stupid move.

Let's say we're not in combat now, we're in a peacetime Air Force—we're in a peacetime military.

Well, it's a little more difficult, but they can tell. It filters down. If you've got a commanding general that doesn't know what he's doing and he issues an order, it's going to be obeyed, yeah. But as it comes down, you can see the reaction of the guy who gets the order. Even the most stupid private can see that and knows what's going on. And they've got to get pretty good at reading signals like that. On the other hand, if it makes sense, if it's going to get the job done and make you proud of being in that outfit, that's reflected right down. There's a feeling.

Is the Air Force going to operate just as well without charismatic leaders, effective leaders?

It'll operate just as well as it wants. Momentum will take them along for a while. For instance, they're still doing things in SAC the way I set them up back in 1950, but it won't go on forever. You've got to get some new blood and some new sparks and some new leadership into the outfit.

General William J. Livsey

In his 35 year career General Bill Livsey earned a reputation as one of the Army's most distinguished post-war leaders. Born on June 8, 1931, in Clarkston, Georgia, he earned an undergraduate degree from North Georgia College and a commission from ROTC in 1952. He commanded every size unit from platoon to brigade, including a battalion of the 4th Infantry Division in Vietnam. From August 1972 to July 1974 Livsey was executive to General Creighton W. Abrams, Chief of Staff of the Army, whom he had served as aide-de-camp when Abrams was commanding general of the 3rd Armored Division in the early Sixties. Promoted to brigadier general in March 1975, Livsey subsequently commanded the 8th Infantry Division, VII Corps and the United Nations Command in Korea, his four-star assignment from May 1984 through June 1987, when he retired. He resides near Atlanta, Georgia.

13

"BE BRAVE-SMART AND NOT BRAVE-DUMB"

A Conversation with General William J. Livsey, USA
June 30, 1992

You certainly have an easy-going style of leadership.

I capitalize intentionally on the country-boy-come-to-town approach. I used it as a technique. I knew I had a Southern accent, I wanted to speak better English all the time. My academic credentials are pretty damn good. I was valedictorian of my high school class, a small school out here in a small town in Georgia. I was second in my college class at North Georgia College. I was never outside the first ten in any military course I ever went to and I was the honor graduate at the Command and General Staff College—Gorman and all the rest of those bright guys were in that class. And I made straight "A"s when I went to graduate school at Vanderbilt. I've been an over-achiever all my life—a classic case of the runt complex. I cried if I made a B+ because I wanted to make an A. Most of all I wanted the respect of the people I was associated with.

The Army you served in is downsizing, as you well know. What effect will this have on the development of young leaders?

We have put ourselves in a forked tongue position. We said for a good while there, "Go forth and do something. Don't be afraid to make mistakes; we can underwrite a few mistakes out of you second lieutenants and first lieutenants and captains." Now that's really not true. As you downsize now you've got one little crimp—not a bend, just a little crimp in your career pattern—and you're out. That's what's bothering those young people out there. God, they're so good. They're much better than we were. My son is a major; he was Colin Powell's aide up until the time he went to the Command and General Staff College. He's not as smart as I am—IQ wise—but I think he's just so much better than I was in terms of dedication. He's got a great sense of humor and I hope he keeps that.

Did you find it difficult as an ROTC graduate to deal with West Pointers in a downsizing Army?

I never had that problem initially. I don't think it's important where a guy went to school when you start leading American soldiers. I was never happy with the negative motivation that characterized West Point. They have tried to change some of that because American soldiers do not respond to negative motivation. If you want to get a guy to stand up in the middle of withering fire all you have to do is tell him, 'Son, I've been watching you and you're one hell of a man. You're something.' They'll go anywhere, do anything—they're the best in the world.

General, what leadership lessons did you learn from commanding a battalion in combat during the Vietnam War?

I know generals, and I won't name them by name, who in my opinion got their stars and their promotions by getting American soldiers killed. I don't like that. I don't like it at all. I think you need to be brave-smart and not brave-dumb and I think the last dumb thing that an infantryman wants to do is to assault the hill. That's the last dumb act you want to put yourself into. You want to use all of your combat power in such a way that the assault is never necessary and that you can walk up onto the hill at the right time and pick up the pieces. That was my philosophy. My battalion in Vietnam fought that way. We had very few soldiers killed—I think we had 16 killed during the six months I was a battalion commander, four of those were killed when a truck hit a mine in a rear area—I know guys who had 90 to 100 killed in an hour. That's one of the philosophies deep in my heart.

The other one is medals. There's a great deal of difference in medals and I can get sworn statements, if you want to, from guys who were in certain units who will tell you there was a package they gave most of their battalion commanders. That wasn't true in the 4th Division under Major General Ray Peers, who is now dead. He made me president of the Battle Board and I was the G-3. I didn't have time to be president of the Battle Board but he said, "You're the G-3 and you know what goes on in this division and I don't want anybody to get a medal who doesn't deserve it." We recommended four Distinguished Service Crosses during that time and the highest rank was a second lieutenant. I tried my best during Grenada, during Panama and during Desert Storm to say to my Army—and it is my Army—please don't cheapen those awards. Please don't do it. Particularly don't give out that combat infantryman's badge. When I got mine, you had to be in combat for 30 days. I have some strong feelings about that.

I understand that command of a battalion is absolutely essential if an officer aspires to become a general.

We knew that if we didn't get a battalion command in combat we were probably not going to be competitive. I knew a guy, one of the outstanding West Pointers—class of 1948 or 1949—who was leading his group in everything—early promotions. During the Korean War he was somehow one of the few who got sent to Europe. In spite of that he had done extremely well. He was a lieutenant colonel and below the zone for colonel before he was even assigned to Vietnam. He told his wife, before he went to Vietnam, "I can't do it." She said, "You're gonna be just fine." He went over there and took that battalion and within two months he was a wreck. They had to relieve him.

What was it like as a young lieutenant and captain during the Eisenhower years, when Army leadership was often found wanting?

I went to Europe and found it much the same way—the good part of the Army. The bad part was, that some of the bad leadership in the middle that we had in the Korean War was still around in the mid to late Fifties. In Germany I had two battalion commanders, when I was a company commander and a battalion staff officer, and neither one of them was worth a flip. Neither one of them. The regimental commander was pretty good but those guys were terrible—no inspiration, afraid that the colonel was going to get angry with them. No confidence—you've gotta exude confidence and your soldiers will know it right away. When I went over there I was a staff officer initially, then they were going to give me C Company, the best company in the battalion. I was delighted. At the last minute they changed and gave me D Company, the worst company. It was terrible. Unfortunately, when I went down there I found out that it was all the company commander—a West Pointer, by the way. He had five of the finest master sergeants I'd ever seen. All he needed to do was turn them loose.

I said to the First Sergeant, "We're gonna do something around this place. First of all, we've got a division softball league, and I want to enter it."

"We don't have time for that, sir."

"Yes, we do," I said, "but this is fast pitch and you can't play fast pitch softball unless you've got a pitcher. Next formation see if we've got a pitcher."

The First Sergeant got in front of the company and said, "This company's gonna have a softball team. Now who can pitch softball?" Nothing. "I said already, this company's gonna have a softball team. Is there a softball pitcher?" Nothing. The third time he said, "Darn it, there's gotta be somebody in this company that can pitch softball." This six-foot-eight private raised his hand and said, "First Sergeant, I'm a softball pitcher.

The First Sergeant said, 'Lockhart, why didn't you raise your hand when I first asked?'

"Cause there's nobody in this company who can catch me."

"You wait right here, Lockhart," I said and I went to the trunk of my car to get a mitt. I used to catch in the industrial league in Georgia. I stepped off the 45 feet there and he threw three pitches. I smiled at him and he smiled at me and I said, "We're gonna win it all." He was a semi-pro from Chattanooga.

You became Commandant of the Infantry School fifty years after General George Marshall was there. A bunch of very capable officers, like future Chief of Staff Shy Meyer and future four-stars Bill Richardson and Dick Cavazos were there at Benning with you.

These were guys who'd been shot at in the Korean War, genuine heroes who became the leaders of the Army in Shy's time and my time, they were like the same bunch of guys down at Benning in the 30s when Marshall was down there. But when I went to Fort Benning to command it that kind of talent didn't exist and I had a hell of a time trying to get it. We used to have Congressional Medal of Honor winners teaching leadership. Now they have some A.G. captain teaching leadership because he has a political science degree. Shoot, I didn't want that. He was a bright young man, but I didn't want anything about Pavlovian Theory—I wanted somebody to get up there and talk about why soldiers go up hills. Fred Mahaffey and I worked awfully hard on that.

I was told to go down and take over Fort Benning and see what I could do with it. That's all I was told. I went down there and took over from a guy named Will Latham, a good soldier who had a propensity to make people unhappy. I was instantly loved. It was easy. It's easy to do things right. If you know that you're going to have certain training exercises with a mechanized battalion you know they've gotta have a tank company. That all has to be laid out. You don't go to a tank battalion ten days before you're going on an exercise and say by the way, I need a tank company. That's not how you do business. Still today, in this wonderful Army we've got, we still have guys who think that you have to do things the same way. You're allotted so many tank rounds per tank per year—it used to be 200, but it's considerably less than that now. We've got dumb guys that think that each tank fires 200 rounds. What you do is you take that pot of ammunition and you look and see that the crew on Alpha 23 shot distinguished last time. It's got the same tank commander, the same gunner, the same loader—all this. Hey, you say, I saw how good you guys were. I want you to show me if you're still that good. Here's a ten-round confirming course. Let me see—boom, boom, boom. Yeah, you're just as good as I thought you were. You can go on home. Alpha 24, new tank commander, maybe never been in there before, a new loader, take some of that ammunition you saved off of Alpha 23 and bring him up to the same standards. You've got guys who still think every time you go down to the tank range every tank oughta fire the same number of rounds. That's crazy.

They used to divide all the talent in the officer corps into thirds: upper third, middle third, lower third. Ostensibly, somehow there were certain assignments—Army staff, West Point, and a few other places—that got five from the upper third, five from the middle third and none from the lower third. When I took over Benning, I was getting 2-2-6, two from the upper third, two from the middle third and six from the lower third. I used to say the strength of our Army, as best I can tell, has always been in its school house: its Fort Bennings, Fort Sills, Fort Knoxes and its Leavenworths and keep them strong. I guarantee you today that if you go to Leavenworth and you see who is on the platform out there you don't have the dynamic, inspirational brigade commanders, the battalion commanders who fought in Desert Storm. They're not there. They're getting their ticket punched in the Pentagon and other places like that. It's a problem—a real problem—because you want talent at West Point, you want talent in the Pentagon. I know Desert Storm was only a test, a four or five day operation, but let them put some of those guys on the faculty at the Command and General Staff College. As the Army downsizes, there are two things that need to happen—and they're sorta in competition for resources.

General Livsey, you can't have spent all the time you did with General Abrams and not have some of that tremendous leader, one of the Army's "great captains," rub off on you.

The first time I met General Abrams he was a brigadier general and assistant division commander of the 3rd Armored Division when I was company commander of D Company, 2nd Armored Rifle Battalion, 36th Infantry—mechanized infantry, what we called armored rifle back in those days. We were down at Graffenwier, one of the training areas you've heard people talking about, and we had just finished an armored mech attack and were reorganizing. All of a sudden this jeep drives up with a big red star on the front—one of the few generals I'd seen in my life. I remember seeing one at a reception line at Fort Benning one time. I went running up to the side of the jeep and stood at attention. He asked me questions for 45 minutes, and I must tell you we had one of the best companies I'd ever seen because I had the best noncommissioned officers I'd ever seen, and I knew what they'd taught me. Everything he asked me I answered bam, bam, bam. As he was just about to leave, he said, "Well, Livsey, that is your name, isn't it?"

I said, "Yes, sir, it is."

"I like everything I see here except that tank over there is about ten yards too far forward." The tank had its engine running, it had a hand-held radio in it, and the First Sergeant was standing where he could hear Abrams. Very quickly when he heard him he simply said, "Two three back one zero." Abrams had hardly got it out of his mouth till the tank backed up ten yards. He said, "'I'll be" and drove off.

I was down at Grafenwier, training as usual, and I got a call from the chief of staff. "Come back to Frankfurt, change clothes and go see General Abrams. I don't know what he wants." I went back and put on my good uniform and went to Heidelberg. His secretary, a lovely lady named Annette Holliman, said he'll see you in a few minutes. I went in there and he was sitting behind the desk. He looked up at me and said, "Livsey, you want to be my aide?"

I was scared to death. I didn't know what to say and I said, "Yes, sir."

He said, "You're it." I was standing there and he looked up at me with this sort of disgusted look on his face and he said, "That's all." That was the last bit of guidance I got.

Most of the time he called me Livsey. Even the day that I left him, and I was a captain, I went in to see him.

"Sir, it's time for me to go."

"Yeah, I know that. Got anything else for me?" I said, "No, sir."

He said, "Well, thanks, Livsey."

The summer of 1972 General Abrams became Chief of Staff while you were commander of the 2nd Brigade of the 4th Infantry

Division at Fort Carson. You had been a colonel only a couple of months. Were you surprised to hear from Abrams again?

I was out in the field in the operations center when the phone rang. The ops sergeant wasn't there at the moment so I ran to answer it.

"Livsey," I answered.

"Abrams," was the reply. "I've got a job I want you to do. What are you doing out there?"

"Sir," I said, "I'm about to wind up my brigade command and I'm going to be chief of staff of the division."

"Yeah," Abrams said. "Now, about that job I want you to do."

"Sir, when should I come there?"

"You take care of the details." Then Abrams hung up.

I took a helicopter back to division headquarters to inform the commanding general of the division, a man of measured patience.

"Sir, I don't know how to tell you this, but I'm going to have to turn over my brigade and go back to Washington, probably next week."

"You little jackass, you're not going anywhere," the general said. "Nobody messes with my officers. You get your fanny back there to the training area and I don't want to hear any more out of you."

I said "Yes, sir," and left. About two hours later, back in the training area, a helicopter suddenly showed up. It was the commanding general. I went over, saluted, and the general said, "Bill, you're going to Washington next week."

The following Sunday I arrived in Washington, checked into the BOQ, and the next morning was sitting behind the executive's desk at 6:30 in the morning. Later Abrams came in, said "Good morning, Colonel" and that was that. He had great faith in the people he knew. He didn't like to break in new people. He knew I'd do what was right and be intensely loyal.

After almost two years as executive to the Chief of Staff, you received orders to return to Fort Carson. Knowing that General Abrams had cancer, it must have been a difficult good-bye for you.

When he was dying I went over to Quarters One and he was sitting in the living room. He was on one side of the coffee table and I was on the other.

"Sir, it's time for me to go now."

"I know that," he said. "Do you have anything else for me?"

I said, "No, sir."

He said, "Well, thanks Livsey."

He got up and I got up to shake hands and he brushed my hand aside and hugged my neck. He started crying and I started crying and I didn't know what to do, so I just stood him up straight, saluted and ran out the back door and collapsed.

General William F. McKee

William Fulton McKee was born in Chilhowie, Virginia, on October 17, 1906. Following graduation from the U.S. Military Academy in 1929 he served in a number of assignments within the Coast Artillery, earning promotion to lieutenant colonel shortly after World War II began. Promoted to brigadier general in February 1945, he eventually earned a fourth star as commander of the Air Force Logistics Command in August 1961. The following year McKee became the first Vice Chief of Staff who was not a pilot. He retired on July 31, 1964, served as Administrator of the FAA, and with General Bernard A. Schriever formed a successful Washington-based consulting firm. He died in San Antonio on February 28, 1987.

14

"A GOOD COMMANDER IS ALSO A DAMNED GOOD MANAGER"

A conversation with General William F. McKee, USAF October 18, 1974

General McKee, your style of leadership, the way you go about your job, would you say that you're more of a commander or a manager? Or is it possible to make a distinction?

I don't think you can make any particular distinction between a commander and a manager. A good commander is also a damned good manager. He has to be to be a good commander. On the other side, if you want to be a good manager—all too often we don't have as many as we would like—you've got to be a good commander and that's where a lot of managers fall down because they fail in the command business. It's largely a business of handling people. The ability to make decisions. The ability to understand people. The ability to get the most out of them. This is true whether you're military or civilian. I don't see a hell of a lot of difference. We see fundamental mistakes. I think one of the great mistakes in leadership was shown by President Nixon. The great mistake in leadership that he made was an inability to select people, an inability to get people who had enough guts to stand up and tell him what the hell they thought, instead of having people around him that told him what they thought Mr. Nixon wanted to hear. I think this is important. I'm now in a very sensitive area, but I think this will go down in history as a major mistake on the part of the President.

Since you're talking about commanders, I'm talking about commanders too. The greatest single fault, both in industry and government, is the failure of people up and down the line to tell their leaders, their bosses, or their commanders or their managers, what they ought to hear instead of telling—which they all too frequently do—what they think the manager wants to hear. I saw this all too often. I think this is a fundamental failure in the operation of our industry today and particularly in government.

Does that mean, General, in your Air Force career you had people working for you who were not afraid to tell you when you were wrong?

I'll tell you, I'll sum all this up by a short story. In early 1962 I was commander of the Air Force Logistics Command. I was the only four-star general in the Air Force that was not a pilot. I was on my way back from Europe on an inspection tour. I got word in Bermuda that General LeMay wanted to see me in Washington—please stop by tomorrow morning on your way to Wright-Patterson AFB. This is a story that's never been written. I couldn't divine what General LeMay wanted to talk about. Didn't have the foggiest notion why it was important that he see me. So I got to Washington and went over to General LeMay's office, he was Chief of Staff. I had a nickname—Bozo—which everybody called me, including the President.

LeMay said, "Bozo, do you want to be Vice Chief of Staff?"

I said, "No."

He said, "Why not?"

I said, "I've got one of the best jobs in the Air Force. I'm pretty much my own boss. I spent many, many years in the Pentagon. Why in the hell do I want to come into my last two years in the service and work for you seven days a week, around the clock, and take this beating? I have no interest in being Vice Chief."

He said, "Now that we've heard that bull, let's go in and tell the Secretary you'll report for duty July first."

And it wasn't bull either, it was true. Before I said I'd do it, I told LeMay we had to have one clear-cut understanding, and this is important in management. I said, "Curt, I have spent years here, as you know. I was Assistant Vice Chief of Staff for six years. I worked under several Chiefs and Vice Chiefs. The only way this office can operate is to be one office. You can't have an office of the Chief, an office of the Vice Chief, and an office of the Assistant Vice Chief. This is the Chief of Staff's office. You and I have to fight our problems out alone between the two of us. When we are in a staff meeting or in the Secretary's office or whatnot, it is absolutely wrong for you and me to take strong issue with each other. We ought to fight these issues out alone."

He said, "I couldn't agree more."

Well, after working there for I guess two or three months, I walked into LeMay's office one day and I said, "LeMay, you're in trouble." And he said, "What?" I said, "You've got a bunch of generals up and down the E-ring here who are busy trying to figure out what you want to hear, instead of what you ought to hear." He said, "Who?" I said, "All of them." He said, "What do I do?" I said, "Tomorrow morning you come into my staff meeting—we have a staff meeting every morning at 8:30—sit down in my seat and give the following speech." And I gave him a fine five-minute speech to straighten the boys out. Next morning he came in chewing his cigar, looked up and down the table at all these lieutenant generals. He said, "I understand some of you guys are sitting down at your desk down the E-ring here trying to figure out what the hell I want to hear. I want to make one thing clear. If I catch any of you doing it, you're fired. Do you understand?" And he got up and walked out. So he condensed my five-minute speech into thirty seconds. But this is the reason that I tell you this, and whatever use you want to make of it. I think he is one of the great figures.

Another great figure is the one-man operation—where one man thinks he knows all the answers. And every now and then you've got people in the military or in civilian life or in government, you get people who think they know all the answers. They're willing to give people responsibilities, but they're unwilling to give those people the authority to discharge the responsibility with which they've been charged. I'll tell you a famous story. I'm going to cover all this management thing in about two minutes and we can be through.

General Spaatz was the first Chief of Staff of the Air Force. Great guy. Just died about a month or so ago. I was one of the honorary pallbearers. It was a Saturday morning, 11 o'clock. I was Assistant Vice Chief of Staff. General Vandenberg was the Vice Chief. He was out of town. I had three papers that had to be signed by the Chief of Staff so I went in to see General Spaatz.

I said, "Sir, I have three papers here that you have to sign before you leave today."

He looked at me and he said, "Bozo, didn't you get promoted recently?"

I said, "Yes, sir."

"Who promoted you?"

I said, "You did, sir."

"Why did I promote you?"

"I don't know, sir."

He said, "I promoted you to take care of papers like this. Do any one of these three papers have anything to do with a war starting tomorrow or the next day?'

I said, "No, sir."

He said, "You sign them. I further think now, Bozo, I just want to be straight with you. I'll let you make one mistake, but if you make two, you're fired. And besides that I'm late. I'm due down at Halls in fifteen minutes and I've got to make the martinis."

I went back. I read those papers about eight times because I had to sign them. And that makes a hell of a difference. But Spaatz was a great manager. He had the tremendous capability of selecting, which Nixon didn't have, selecting the people to do the job. Give them the responsibility and give them all the authority that he had to discharge that responsibility. LeMay was the same way. That's the reason LeMay was a great leader. LeMay used to call, every month, every new colonel in, and he'd call him into my office later on when I was Vice Chief under LeMay. And he made a very short speech. He said, "I spend seventy-five percent of my time either with the Joint Chiefs of Staff or testifying on the hill or on other matters external to the Air Force. I only have twenty-five percent of my time to devote to the running of the Air Force. Gentlemen, General McKee is running the Air Force. I would like to give him the complete responsibility but under the law I have it. But I want you to know that he has every bit of authority that I can give him to carry out his job. Now if any of you guys don't like the way General McKee is running the Air Force, my door's open if you can ever catch me." But this was an understanding and I felt it very deeply. I didn't have to wait and go in and ask General LeMay every time I made a decision. I kept him abreast, I kept him advised of what I was doing, and we had an absolutely perfect team. This is where you have understanding between people and have mutual trust between people. This is what makes things go.

Did he ever question any decision that you made?

No. Oh, I should take that back. Sure, all in very good nature, sometimes yeah. He would question if I was sure I was right, but we never had any major boo-boos. Nothing fundamental. If we had a fundamental thing of very great importance to the Air Force or to DOD or to the country, heavens, I would always discuss it with him. I wouldn't arbitrarily go off on my own because it was too important. The smaller things didn't make much difference one way or the other. It didn't bother him. I mean, a lot of little things some people get mad about. He never bothered with those. He'd let me take care of them and he didn't have any problems.

General, did you, in turn, pass along the ones that you felt other people below you were competent to handle? Was this the way it worked?

I'm probably one of the leaders in this whole business with all the years I spent in the Pentagon building in that philosophy. I carried this philosophy on when I became administrator of the FAA even. President Johnson asked me to take over that job after I retired. This was kind of hard to get through to FAA. They hadn't been used to this. But I called all the head guys together and pointed my finger at them and said, "Look, I'm charging you with this responsibility and I want you to be sure to understand what your responsibility is. And I'm giving you the authority to do it. And if you can't hack it, my friend, I'm going to get somebody else." This was something new to them.

So in your style, General, there was no room for mediocrity?

You had to move them out and it's very difficult. There's still some guys over in FAA that I had to move or fire and I still see them occasionally around town and they look at me and say, "Son of a bitch." But you have to step up and do it. How else can you run a big job? This was a great thing about General Arnold. General Arnold was one of these really great leaders when the war started. He had all of these contemporaries of his around and they got promoted. God knows how many dozens of them he had to fire and get retired. They couldn't hack it. And the old man, I think this ended up causing his heart attack which he had in 1945. Basically, he was pretty soft hearted but he knew he couldn't get his job done without doing that. He had to get younger people in. He had to get people in who could work twenty-four hours a day, practically seven days a week. General Arnold was one of the big leaders of this whole century as far as I'm concerned. He was as much responsible for the U.S. winning the war when we did, or more so, I think, than any individual. He was the guy who finally got through to General Marshall and to General Eisenhower in Europe the importance of air power to winning the war. And they believed him. This changed the whole course of things.

Was there ever a doubt in your mind at the time that you were commissioned that you would eventually become a general officer?

When I was first commissioned I never dreamed about being a general officer.

It wasn't one of your goals?

Listen, when I was commissioned—that was 1929—your goal in those days was you wondered if you'd ever get to be a captain, much less a general. A general was like God. As a second lieutenant you never saw one. Second lieutenants in those days, they never dreamed about being generals. One of my relatives was a brigadier general on my mother's side. You'll see his picture hanging on the wall. My mother said, "Son, maybe someday you'll be a general." And I looked at her to see if she'd lost her mind. Unfortunately, she didn't live to get to see me be one. My father, a medical doctor, died at ninety-three. He was able to live through my getting to be a four-star general and being administrator of the FAA. He got a great kick out of it.

I can well imagine. General, can you relate, please sir, the circumstances and the events concerned with your selection to be four-star rank?

I can tell a story before then that I think was more important as far as the circumstances were concerned when I got promoted to brigadier general. It was in 1945. I was Deputy Assistant Chief of Air Staff Operations, Commitments and Requirements. Someone walked in my office and said, "Your name is on a list that has just been sent up the White House to be a brigadier general." Of course I got pretty puffed up about that. So I picked up the phone after the guy left. As a matter of fact it was Dr. Edmund P. Learned. He was a great guy. I don't know where he found out about it. I called up my wife and said, "Honey, your husband has just been nominated by the President of the United States for promotion to brigadier general." I expected her to fall back and swoon from the telephone. And she said, "Dear, I think that's great. Please stop by the market at Shirlington on the way home and get two dozen oranges. The boys have no fresh oranges for orange juice." So I punctured the balloon and it stayed punctured ever since. When I got to be a four-star general there was no balloon. No, I was the vice commander for a number of years for the Air Materiel Command under General Ed Rawlings. I was a lieutenant general. Rawlings retired, I think it was 1960, and General Sammy Anderson came out and was given his job and got his fourth star. At the end of about a year there was a change going on in Washington. General LeMay had been Vice Chief of Staff and he was about to be made Chief. It hadn't been announced yet. He called me one day and said, "Bozo, I think I'm going to be Chief and if I am, I'm thinking about putting you in charge of the Air Force Logistics Command. Do you think you can hack it?" And I said, "Yes, sir." So that's the way I got the word. And in two days he was named Chief, Sam Anderson was transferred to Europe and I was made commander.

General, in your dealings with people throughout your career, how did you reward those who did an extra effort for you, put forth more than what was required of them?

Only two ways you can reward those people. One, you reward them through the medium of their efficiency report because it's through the efficiency report system that they get promoted. Two, is putting them in the most responsible job for their rank that you're able to do. For example, my bosses over the years did that to me and many times as a captain, as example, I was in command of a regiment before I was transferred to the Air Force in Puerto Rico. I was the only captain in the whole U.S. Army to command a regiment. In the Pentagon building I successively had jobs calling for one to two ranks higher than the rank I had. That's how you reward people that you think have a lot of potential, a lot of ability. Maybe they made a big mistake. I didn't have it, but at least they must have thought I did.

In other words, you never were inclined to some of the things like sending a guy a bottle of whiskey and having it on his desk when he worked late or something like that?

Never even dreamed of it. Never. Nothing. This is absolutely totally foreign to the military system. I never did it out in civilian life when I was running the FAA. When a guy was doing a good job, you call him in and tell him so. Let his colleagues and contemporaries know he was doing a good job. Of course, in the military you also have the awards system where the guy who really does something outstanding can be given the Legion of Merit for doing an outstanding job; or if he was a senior general with great responsibilities, you might give the Distinguished Service Medal. I have three. One of them was pinned on by General Arnold, one of them was pinned on by General Vandenberg, who was then Chief of Staff. The last one was pinned on by the President, LBJ, when I retired.

Sir, someone told me the story some time ago that LeMay showed up late one evening at a remote SAC base somewhere and the MP, recognizing who it was, waved him through without stopping him

to ask for an ID. LeMay went back and chewed out the MP for not firing at the car when it went through the gate. **Is it possible this is true?**

I never heard that particular story, but so many of those stories go on, I can't verify that. I don't know if that's right or not. LeMay was a very hard taskmaster. LeMay did a fantastic job in SAC. After all, he kept SAC honed at a razor edge there for all those years when we weren't fighting. And how he did it, very few people understand. As McNamara told me one day, and McNamara and LeMay didn't get along very well, but he told me, "Bozo, I will have to say if we do have to go to war, I want to be on LeMay's side". That's a great compliment. LeMay's a great combat leader and he was also a great leader in peacetime. He was a man more responsible than anybody else in all the services for getting decent housing for our troops. He led that battle, he led that fight and won it when other people were content to sit around and let the enlisted personnel live in barracks, World War II type barracks.

He has the reputation of being so tough and playing hardball, actually he was pretty soft hearted. The troops knew that he was in there fighting for them. This makes a hell of a lot of difference. Now I'll tell you another story, a story about LeMay. This has to do with the Cuban Missile Crisis. This one's never been told either. During the Cuban Missile Crisis, LeMay, when it happened, and I was deeply involved with it, LeMay was in Europe on a trip. So all that week I was sitting in as a member of the Joint Chiefs representing LeMay. I was acting Chief. When the book came out about Robert Kennedy, at least it was reported in the paper—I haven't read the book—the statement was made that during this period LeMay recommended dropping the A-bomb on Cuba. You may remember this. LeMay wasn't even there. He got back the last day. He wasn't even there. This is what burns you up. Robert Kennedy got a million dollars for his book about Cuba and all my notes that I made at that time are in a top secret file in the Pentagon building. If I'd written all this I don't know where the hell I'd be—probably be in jail. But these are the kinds of stories that go on and I'm actually telling you the truth.

You were on particularly good terms with McNamara. When the administrations changed, and Melvin Laird became Secretary of Defense, did you see a perceptible difference in the way they conducted business in the Pentagon?

Oh, yes, very definitely. That makes the point that you want to make. There was a real difference. Now when Laird took over as Secretary of Defense there was very excellent rapport between Laird and the chiefs. And really it's to get the damn job done. The Secretary of Defense can be a rough, tough guy but there's got to be mutual trust and understanding. After all, the military guys aren't in there to try to wreck the country or try to put the country into financial straits just because they think they've got to have a big Army and a big Navy and a big Air Force. Actually, this is something I don't think is understood by the great American public. Most senior officers, admirals, generals, all of them, if they really felt in their hearts and souls that by taking off their uniform down here at 14th and F, and buying themselves a pair of pants to go home in, it would stop the war and we'd never have another war—they'd all do it. Because most of them have seen wars and they don't want any part of it. They know the horrors of war, they don't want it. They know what it does. On the other hand, they've also seen what happened to Europe when the Nazis moved in. Go back throughout history and see what's happened and none of us want to see that happen to us. You see, we're spoiled. We've never been invaded. We've never had bombs dropped over here.

But you go to Europe and see what's happened in Europe. A year ago last September, my wife and I went to Russia, went to Poland. You see what happened in World War II. Warsaw was absolutely demolished. You go to Leningrad and see what happened there with the siege of Leningrad that went on for months and months with people starving to death and living with absolutely no heat in subzero temperature. These people have been through it. We've never even been through this. I hold no love for the Russians whatsoever, but you can well understand why the Russians insist on a divided Germany, because they've been run over too damn many times. It's pretty easy to understand when you see what happened to them.

General, how did you relax and take some of the tension off of yourself?

In the Air Force?

Yes, sir.

The best way, like I did in World War II. A couple of times I got routed, going home late at night, and happened to pass by Arlington Cemetery and looked up and said, "McKee, don't be so stupid. Look at all those indispensable people up there." So I'd go home and have a couple of drinks and felt like a new man. Got so that about every six months, when I thought I was pretty damn important or whatnot and had to work fourteen hours a day, seven days a week, I'd go by Arlington Cemetery and look at all those indispensable people and realize that I was just as dispensable as they were. Most of those top-flight guys at one time thought they were indispensable or somebody else thought they were. But there wasn't any relaxing during the war. You just worked seven days a week. As a matter of fact I had two little boys living over in Fairfax. Finally my wife said I had to come home some Sunday afternoon in the daylight so she could walk me around Greenway Place so the neighbors would realize that she had a husband—which I did one day. You know, I never even met any of these people. After that, though, in peacetime, sure, you get a chance to relax, play golf, take some trips, take a little vacation.

One of the hardest jobs I ever had really was as administrator of the FAA. It you take it seriously, that's a seven day a week, twenty-four hour a day job. You're at the end of a phone twenty-four hours a day because you're responsible for the whole air traffic control system of the United States. You're responsible for the safety of the entire system. I used to hate to pick up the telephone after midnight because I knew it was bad news. In this job I don't have any of that tension from crisis. I was 68 years old yesterday, so I'm entitled to a little bit of relaxation.

General Edward C. Meyer

A member of West Point's class of 1951, where he acquired the nickname "Shy," Edward Charles Meyer was born in St. Marys, Pennsylvania, on December 2, 1928. Six months after graduation he was sent into combat in Korea and earned a Silver Star. Following four years in Europe and two more in the office of the Chief of Staff, U.S. Army, he commanded a battalion in Vietnam in 1965-1966 and a brigade 1969-1970. He served as Assistant Commander of the 82nd Airborne Division, was a fellow at Brookings Institution and Commanding General of the 3rd Division 1974-1975. While serving as Deputy Chief of Staff for Operations & Plans in the Pentagon, Meyer was named Army Chief of Staff by President Jimmy Carter. General Meyer retired in 1983 and is active on the board of several major corporations. He resides in Arlington, Virginia.

15

"EVEN IN RETIREMENT MENTORS CONTINUE TO PLAY A ROLE"

A conversation with General Edward C. Meyer, USA
March 24, 1991

It's well known in the Army community that you believed strongly in—and actively practiced—mentorship. How did mentorship affect you in the early years of your career?

First of all, you have to sort of define what the components of mentorship are. One is in the area of guidance, counseling, advice, teaching. How did you learn from that individual, why did that individual take time to teach you? That's one facet of mentorship. The second facet is door-opening; it's providing opportunities for an individual. Mentoring, in practical terms, is all of the things that occur when you get into those two buckets. And that's different than your normal relationship can be with your boss or your commander. Your boss or your commander can handle you as a very competent, very capable individual and tell you what to do and so on without taking the time to teach, advise, counsel or he can do it without feeling an obligation to open the doors for you.

In my case, I'd say the individual who had the greatest impact on me was General Jim Moore when he was chief of staff of SHAPE. I was his aide and assistant executive officer. One of my classmates,

Lou Michael, had been his aide for a couple of years and I had moved over to France and lived in Orleans and had a very good job—a job which I enjoyed—in the G-1 section of the Communications Zone. I had a chance to go play golf when I was home, we liked it there, we had good friends down there. About that time I got a call saying, "General Moore is looking for an aide this summer. Would you like to do it?" And I said, "Well, let me talk to Carol and I'll call back." I was up at Oberammergau at a school and I said, "Screw it, I'm not going to go up to SHAPE and deal up there. I'll just stay here and enjoy what I'm doing." I had a good boss and good people to work with. I got a call from Lou Michael. "Well, what have you decided?" And I said, "I've decided not to do it." As soon as I said that he said, "Wait a minute." Then General Quinn, who had been my battle group commander with the 101st Airborne Division and now worked directly for General Moore, the chief of staff, got on the phone and said, "You get your ass up here. You need to come up here. It's an important job for you." I reluctantly said, "Yes, sir." This highlights the fact that life is never simple. There's always people involved whom you've served under, who therefore have an opinion of you—good or bad—and want you to go on to another position.

So I went up and worked for General Moore for a couple of years as his aide and assistant executive officer. It was a period during which he taught me an area of the Army that I hadn't been exposed to—staff work and international operations. In two years I was intimately involved in every piece of staff work that was developed because he permitted me to do that as a young major. And I did all the traveling when General Moore would go so I got the opportunity to sit in on discussions that he had with heads of state and heads of other armies. He was very good also because he permitted me to deal with and observe General Norstad in the same capacity. I played golf with General Norstad, so I had an opportunity to see him on a personal basis. General Moore's function was principally one of providing the opportunities, to provide the guidance when I screwed up—because I did—of teaching me by example. We would have one-on-one discussions on broad issues on which he and I would disagree. He knew that he was getting the views of someone who wasn't in World War II—talking to a young punk about issues of broad international interest.

I think when you talk about mentoring you have to think of the people who took the time to provide guidance, took the time to teach, took the time to counsel when you needed counseling—and that's asschewing as well as back-patting. He played an important role in opening doors. For example, I went to the Armed Forces Staff College and when I finished there two of the officers who had worked for General Moore at SHAPE brought me in to work at the office of the Chief of Staff to coordinate the Analysis Group. As a result of that relationship, and the relationship with General Corcoran and General Woodward, that provided me the opportunity to work in the office of the Chief of Staff as a major and lieutenant colonel and deal with them, with General Abrams, with General Johnson, General Wheeler—plus deal directly with all of the deputies. Our responsibility in the coordination of the Analysis Group was to go down into the organization and make sure that what the general said was really understood so that troops didn't waste a lot of time developing papers that were responses to what they thought the general wanted. I got to watch the infighting within the Army staff for greater authority and greater responsibility and everything else. All of that was a result of General Moore's mentoring.

Even in retirement mentors continue to play a role. General Moore had been a close associate of General Wheeler and General Johnson and had been respected by most of the senior leadership. So, if he makes an observation or something about you at that point in time then you're more likely to get a job than somebody else is. If you're fortunate enough to have a mentor who is highly respected you're all right. If you have a mentor who's not highly respected you're probably in trouble. There are mentors who aren't highly respected. You can get into an area where the boss has either moral problems or bluster problems. When he's gone there's nobody to pick up afterwards because that person's reputation may be such that the rest of the folks that watch them aren't really anxious to have his style of leadership, or his imprint, continue in the Army.

Using General George Marshall as an example—he was one of the most powerful and influential men the Army has seen—how far into the future did his influence extend, in terms of mentoring?

He started out after World War I to develop a coterie of people in whom he had confidence. He kept a book in which those people's names were recorded so that when jobs came up he checked to see where they were. Once he brought them to Benning with him. Ultimately he brought them back up to Washington into critical positions and then promoted them into key squads in World War II. I would say that mentoring is a personal thing. It doesn't continue so that Marshall's people developed a small group who developed another small group who developed yet another small group. It was totally dependent upon the individual and whether he believed in mentoring. Most people in that era did, but it was not a collegial attempt to foster others. It's basically because it's not possible for the people who are involved to be able to use some past clout, such as Marshall, and they have to mentor in their own right. They can't be mentoring based on the fact that they were a "Marshall man." They have to be mentoring as individuals, teaching as individuals. Certainly they used some of the same techniques that they watched Marshall use, but mentoring is so individual and so stylized that I would find it impossible for anybody to draw a conclusion that there was any sort of group that grew up and continued. I can't think of any of them who weren't so unique in their own way that they suddenly developed a collegial approach.

While we're on the subject of Marshall, one of his strengths was absolute candor. Another was separating the military man from the political animal. Have you seen these traits exhibited by subsequent Chiefs of Staff that you've observed?

If you go back to Marshall's development he had an opportunity to observe Washington first hand with Pershing. He had an opportunity to interface with senior military leaders and the White House and the congress. He also saw, at that time, generals who were very political—some of whom he was in competition with to be Chief of Staff. It was his nature that the best way to run the Army—particularly at a time when storm clouds were brewing and then during the war—was to stay out of politics on a personal basis as much as he could and to provide the type of professional military advice that the President didn't get from all the rest of his political advisors. I would say probably out of all the people that I have watched and observed Marshall was best able to try to be sure to perform what he considered to be his primary function, and that was to provide the best military advice.

Many others got trapped inadvertently or directly by background or by the nature of the President into bringing domestic political issues in too early into military recommendations. The only just and proper role for the senior military is to provide the best military advice he can. Where it conflicts with domestic or international political issues, then to understand that he has to modify that advice. But not to give the best military advice seems to me to be a dereliction of duty.

So, do you fault General Maxwell D. Taylor in this regard?

Do I fault Taylor for not giving direct military advice? What did he do that was not solid, sound military advice? There was not a clear goal and objective, there was not an understanding of the importance of having the ability to replace units over there. We never called up the Guard and Reserves because Johnson said don't do it and Taylor went along with that. Harold K. Johnson tried and lost that argument. He went to President Johnson and told him why the Guard and Reserves should be called up. If they had been we wouldn't have had to go to an individual replacement system over there. We could have replaced with units. We could have brought people back. We could have created new units and we could have done a lot of other things. Once you restricted the size of the Army, to be able to sustain a war forward you built in these enormous anomalies which prevent you from having quality forces. More people died than should have. They had poorer leadership than they should have because we had shake-and-bake NCOs and officers that had a lot less experience—and they didn't know one another. He said he should have resigned over it. Probably was right about that. He should have resigned over it. I guess if you look across the broad spectrum Taylor introduced political issues into military advice earlier than any of his predecessors. Remember, Ridgway came back to focusing on the military side and they got tired of him because he was really hammering on the military side.

Can the same parallel be drawn with the relationship between

General Westmoreland and President Johnson? Things like staying in the White House when he came into Washington?

I think it's very seductive. You start to lose track. I think military leaders who are seduced by the President start to lose track of what their principal responsibility is. Some, I am sure, can handle the relationship adequately, but it certainly tends to seduce you into trying to be more responsive to the President. Everybody that's in a military position was put there by a President, and if it wasn't the President that's currently in office he's out there at the convenience of the President who is there. He can move him out any time he wants. There is a sense of loyalty to the President. Where you may disagree with him, unless it's a matter of morals or major military action where lives are going to be lost, then you tend to give him the benefit of the doubt. I'm a bit more of a hard-liner on that. I believe it's in the best interests of the country to have a senior military less politically oriented and more focused on the military art, but also understanding that economic, political and strategic non-military issues have a big impact on how you go and be able to deal in that milieu. When they sit around the table their principal responsibility is to bring to that discussion the hard, clear facts as they relate to the application of military power—so there is no doubt now, and there is no doubt in the future, about what the military implications of political ramifications are going to be. When they stray from that then I fault them. I fault military leaders when they stray from that narrow construct. I fault them because then they are no better than the civilian hucksters around the Beltway who come in and then go out. They permit themselves to be brought down to a level that they shouldn't permit themselves to be lowered to.

You were assigned to the office of the Chief of Staff back in 1965. Did you witness signs of the evolving political bureaucracy that was growing at the Pentagon and the effect it was having on military leadership?

The only thing I saw, that showed me the beginnings of what was going to be a problem, was that every time a paper came in that had anything to do with Vietnam, it had to be cleared by Joe Califano and by McNamara. It didn't matter whether it had to do with proposing that geese be used as an alert force. Somebody would write a letter and say, have you thought about using geese? Peacocks, they'd say, make lots of noise. Every idea. Even when you had it seven times the response and everything else had to go through Califano, had to go through the Secretary of Defense. To me it was an inside observation of how the Pentagon was beginning to work differently. Up until that time, military strategy, military judgment was the dominant factor in determining what sort of forces we were going to have. If you grow up in our society you understand that there are going to be difficulties, but I would argue that there have never been intrusions by civilians into the heart and soul of the military like there was in McNamara's day, where basic fundamentals of soldiering were challenged and changed on economic grounds without ever understanding or appreciating—or being willing to discuss—the importance of those factors to cohesive, effective units in combat. Systems analysts were brought in who challenged everything on a quantifiable basis, who said you don't need to have as many people if you replace on an individual basis as you do if you replace by units. Therefore you don't need as many units because you can use individual units and it's cheaper. It is cheaper, but people's lives are at stake and they die. There are still a lot of people in the personnel business who know it's cheaper and continue to propose it even though they know it's wrong.

Admiral Paul David Miller

A native of Roanoke, Virginia, Admiral Miller earned a B.A. from Florida State University and an MBA from the University of Georgia. Commissioned into the Navy Reserve, he decided to make the Navy a career and served on active duty for thirty years. Admiral Miller was commander of the Seventh Fleet, Deputy Chief of Naval Operations (Naval Warfare) and Supreme Allied Commander Atlantic, from which he retired in 1994. Thereafter he was active in the private sector. Most recently he was CEO of Alliant Techsystems, a $4.5 billion aerospace and defense company. He spends his time in retirement in Florida and near Williamsburg, Virginia.

16

"I PROBABLY WOULD NOT HAVE STAYED IN THE NAVY IF IT WASN'T FOR ADMIRAL ZUMWALT."

A conversation with Admiral Paul David Miller, USN January 28, 1997

Admiral Miller, can you attribute anything directly to the changes that Admiral Zumwalt enacted—things that affected you?

Yeah, I can attribute a very basic one. I probably would not have stayed in the Navy if it wasn't for Admiral Zumwalt. That's down to the ground floor, right?

Right.

I personally was one who came into the Navy as a reserve officer and I had no preconceived notions that I was going to make the Navy a career. None. I was going to serve and then leave and go to the civilian life, but as I completed my time on active duty—Vietnam was going on—I came back to duty in Washington and was going to transition out of the Navy from Washington as opposed to sea duty and particularly duty over there. I was on a destroyer, the typical Vietnam destroyer operation. There are two things that he did that really got my attention. One, he said youth isn't all bad. Youth is good and we need to motivate and give a certain slice of that youth an opportunity to move forward, to take on more responsibilities than the system would have historically permitted them to do. I was fortunate enough to be on the 0-4 list, the lieutenant commander list, with not much time on active duty and still a reserve officer. I think I was the bottom person on that list for that year. I didn't even know I was eligible. That's how close I was paying attention to it until somebody told me that I'd been selected early for lieutenant commander, which came out to be at least two years early, if not more. And I was able to do it as a reserve. I think those two propositions would have never been able to unfold if it wasn't for someone like Admiral Zumwalt saying give talent a chance. So, that was one which led to the next.

An opportunity came up to command a frigate called the *McCoy*, which had historically been commanded by commanders or senior lieutenant commanders who had done their XO tours on destroyers. I was selected to serve as a CO but I was still a reserve, so I had to augment to do that. I said okay because I was still in my late twenties at that time and to serve as skipper of a frigate was interesting. So I did that and things happened properly. I didn't have any problems. Those are just two examples of opportunities that would not have been there if it wasn't for the guidance and for the perspective of Admiral Zumwalt saying hey, there's lots of talents out there. It comes in lots of varieties. It comes in those who have age and wisdom and it comes in those who have vitality in the mid grades and it comes from those who are the younger ones who have real energy and want to work awful hard to show that they can do what the others can do, except maybe a little differently. He gave people such as myself a chance. I've always looked back on Admiral Zumwalt as looking at those problems that the Navy had at that point in time, dealing with them squarely, and then trying to take the institution of the Navy forward. I signed up to be a part of it and then left it after making four stars even quicker than Admiral Zumwalt did. But if it wasn't for him, I'd have never started—and that's true.

Is it possible to separate tradition from institution when it comes to the Navy? We hear a lot of criticism about traditions being cast

aside. Is tradition being used correctly or is it just the old way of doing things and resistance to change?

I think tradition sometimes gets muddled. Tradition, I think, is something that is not precise. Tradition is something that one can rely on as a basic way of approaching an activity: the tradition of going to sea, of sailors being tough men, getting tasks accomplished, being held responsible—all those attributes which are good. I can see that the tradition of a sailor, for instance, standing a helm watch in 1996 just isn't applicable because we have the technology to keep that ship on course without having a helmsman there, without having, for the first time, a sailor moving that ship back and forth. So, have we broken a tradition by doing this electronically, by putting this bridge on board a ship that does it all automatically? I think not. I think the tradition is still there of a ship at sea, a different way of going about doing the particular tasks, because, you know, it changes that a little bit, but it hasn't broken the tradition. That's why I say traditions sometimes are a bit muddled. Take the uniform changes over the last 217 years. Admiral Zumwalt made his changes. People have tinkered with uniform changes but most people didn't like it because they said it's breaking tradition, but when you look back over the continuum, so what? It doesn't break tradition. It just changes it a little bit.

You said something earlier about the Navy as an institution being unwilling to change.

From my personal perspective it's only a 30-year slice of time. That's one-seventh of the period of time that our Navy's been in existence. So it's a small window but it was a period where technology has permitted more advances than probably in the previous 180 years would have allowed. So when I say that the Navy sort of resists change I think that the mariner—which goes back to B.C.

—the mariner has grown up in such a way that they embrace change probably slower than any other profession. They're independent by nature. They have a hull, they have some power system—whether it be oars or sail, steam or nuclear power—they rely on navigation capabilities from the stars, but now it's a GPS. They embrace change slowly to do what they've historically done on the sea and that's go from point A to point B, and if there's something stopping them from doing that, deal with it en route and they're known as sea battles, or whatever you want to call it. When you have a heritage like that you can see why they embrace change slowly and you become a very independent operating force. The Navy was very independent in our history because of communications. The history of either the Royal Navy admiral getting his orders from the Queen, or our naval folks getting their orders from what was then our National Command Authority, and you don't have any touch with them. They're supposed to go out and do their task, and report back, and it's done, right? So that breeds some independence. Not so with technology that permits others to meddle into things naval and then with us having to do our fighting with other services; that's caused us to look at change across the board very, very carefully. If you look at the record it shows that the Navy embraces those kinds of things.

Can we talk candidly about Mike Boorda and your sense of what he left the Navy with?

I left the Navy soon after Mike Boorda became the Chief and I became a businessman then. I didn't follow very closely the particular policies that were implemented. So I would be as candid as my knowledge would permit me to be but I really can't. I was nonstop on a total other career. I went back to the Pentagon on business for the first time just last month, when I went to the retired four-star session because I'd been totally in the commercial sphere trying to learn that and trying to see what the international economic situation does to our national security side, not the everyday workings of the Navy, so I'm just not in a position to comment. Historically, Mike and I, we grew up professionally sort of together. He was a few years older than I was. You know, we sort of knew each other professionally as we went up.

Well, he had his fans and he had his detractors. Which line did you stand in?

You'd have to be a fan of Mike's because he had the Navy's best interests at heart. He felt a kinship to every sailor. But just because he was an enlisted and others weren't doesn't mean you have a monopoly on that. My father was a seaman—made it all the way up to chief warrant officer, and I feel as close to enlisted people as you can feel to them. As a bosun he'd take me down to ships and turn me over to his first class bosun and I'd chip paint on the deck all summer. I think any senior officer that has had the privilege to command feels close to the enlisted personnel right up through the ranks. Mike was particularly good at articulating and making the sailors feel comfortable, which was very good.

Going back to the period that you had your first command, among the crew, how did they react to the changes that Zumwalt had enacted? Was there a tendency for some to be overzealous in following Zumwalt's lead?

I did not have a problem dealing with the crew or in exercising command. I perceived a problem with my leadership. By that I mean we had to do everything, every evolution, 10% better than all the other ships in the squadron to get equal credit because we were all young. We were a young crew, so we didn't have the seasoning, so to speak, of other ships commanded by commanders in the squadron, so we had to do it better to be as good. We used that as a mechanism to pull the ship tighter together and move forward collectively. I thought that the basic crew fully endorsed the Admiral Zumwalt approach about being aggressive and taking risk and we took it on as a challenge.

As you should have.

That's what's on the leadership side of the coin. That's how we turned it around, whether it be for an inspection or whether it be for going alongside for supplies and being as precise and as sharp as we could, or whether it be the conduct of our crew ashore in a foreign port. We wanted to make sure that we were right at the top of the list when any report card was given.

What percentage of your crew, Admiral, were black?

I don't know because I was race neutral. I can remember being in the boiler room with a B.T. named Blossom. You just reminded me of him. He was a young black petty officer. I don't know how many blacks or whites we had—it didn't make any difference. I can truly report that it absolutely did not make any difference on that ship and it was never a hint of a problem. This was one ship in 1972 while lots of things were going on in the Navy, but in our little world there was not a hint of a problem because we fostered the environment that all of us were pulling together.

Then great credit has to go to you, because that was certainly the exception.

No, the credit goes to the individuals who were serving on that ship. The goal was making that ship the best that we could. There are still people in it today. My communications officer is an admiral on the Joint Staff right now. He's a young one-star. The weapons officer is the Appropriations Committee Liaison officer. The engineering officer is a very renowned lawyer in Philadelphia. The electronics materiel officer started the stock exchange in Zambia. He is an international stock expert. We had people that pulled together, and I think it's amazing that I can cite those five guys right there, now 25 years later.

Can we not attribute that directly to the fact that you were identified as talented and were given that opportunity early?

I think you can attribute it not necessarily to my being talented but there was some recognition that I had potential by some authority, and then giving potential the opportunity to either produce or to fail. In this case we happened to produce. I'm sure the record shows that some good people fail, but if you identify the talent carefully enough, my instinct—as Admiral Zumwalt's was—is that the majority of them

will reach out and clear that bar regardless of how high you put it.
Now I, as skipper of that ship, during that time, made sure that we were cultivating an environment where those things would take place. My instinct was that if everybody did that there wouldn't be a problem overall. I mean, that's just a natural extension, right? I didn't step back because I was "head down and blank up" working that ship on a daily basis, right? I didn't step back.

Well, what about after that as your career progressed?

As the career progressed I had the good fortune of working in Washington with three super vice chiefs and in that office you see all the good and all the bad that comes across that desk that has to be adjudicated in some fashion—even more so than reaches the CNO's desk because, you know, a good vice stops a lot of it, right? One of the vice chiefs was a close colleague, Worth Bagley. I started with him. Then I became attuned to problem areas, whether it be in operations or acquisition, or human resources or whatever. Then I went to the Guided Missile Destroyer Command and I had a much broader perspective. I looked for lots of other things that might be going on and felt that I was a better skipper the second time around. For sure. Then as I became a flag officer I knew I was better because I knew what to deal with and what to let go. That's the key of leadership, knowing what your troops could take care of without help and you spending your time on things that they need help with. That's my personal recipe and I've transitioned that to business.

I read recently that the quality of Navy people has continued to go up and is actually at an all-time high. Why?

Because of the women. Women have skewed this because of the quality of their job performance, of their willingness to work with one another, all those reasons.

I share this with you to ask if you'll share with me your thoughts about the increasing role of women in the Navy. Whether there should be one. Or are we going in the wrong direction? As you well

know, critics decry that the Navy is simply doing a social experiment with the increased employment of women.

It's hardly a social experiment. It's something that's a reality. I think we need to groom that new breed of people, or at least the leadership in our own forces. I can recall a poignant reminder inflected in a bit of graffiti on the broken Berlin Wall. The wall is down but how long will it take for the wall to come down in our minds? Any revolution in military affairs will come to nothing if there is not a revolution in the minds of the people charged with understanding it—the balance between commissioned and noncommissioned officers, promotion and retirement policies, and structure of individual units. Flexibility and imagination will need to become more commonplace. The talent problem will be closely linked to the problems the service has faced in coming to terms with military forces that reflect the makeup of our society. We understand this about men and women but have we come to terms appropriately with other aspects? Participation by the wealthy class as well as the underclass, the physically disqualified but intellectually brilliant, and by foreign experts as well as American citizens.

It's more than gender. Technology now allows us to look at it differently. The physically disqualified I mentioned there. Why shouldn't we have people in our armed forces that may have lost a limb in a bicycle accident as a youngster but he's intellectually brilliant? Why shouldn't he be in our armed forces today, when he can do a task that has nothing at all to do with the foxhole or with some other attribute for what we used to look at for military service? We have to start looking at those kinds of things. Admiral Zumwalt would clearly be looking at those kinds of things, I believe.

Well, why is no one else?

I don't know. Because how did we start this?

I think you're right.

I rest my case. I believe that there's those kinds of people. It's ruled the system for centuries. Admiral Zumwalt, as I understand the legend, would have never risen to that position as quickly as he did had it not been for political sponsorship, right?

There hasn't been leadership, civilian political leadership, that's reached down and said hey, I've combed the list of the talent pool here and it's time for someone that's not quite right. Right?

I think you've raised a very critical point, Admiral, that I want to carry forward. That is without the inspired leadership and vision of the civilians, these people are not going to get anywhere.

That's right. They can move forward within the service but for them to move forward to positions of great responsibility then they have to have some backing by other than uniformed membership.

How can we improve that situation? What do we need to do as a country?

I think the best way is within the overall national security in the national command authority apparatus, which is having leaders there that understand and appreciate all factors of the national security equation and are comfortable with having those that are out on the point of the spear working for them.

So what does that mean for the Navy as we get into the next millennium?

For the Navy, they need to help the nation in deciding one of two things. Are we now, because of the change in the international security environment, a continental based force that will project power as required, or are we that maritime nation with global interests? This is a big different set of checkers on how you do things. I personally believe that we're a combination of both and what the Navy must do, along with the other services, must meld together the strategy that permits us to be a land based power but also allows us to be part of the security environment without becoming part of the background noise.

Admiral Thomas H. Moorer

One of the Navy's most distinguished leaders, Thomas Hinman Moorer was born on February 9, 1912, in Mount Willing, Alabama. Following graduation from the Naval Academy in 1933 he served aboard two battleships and then earned the wings of a naval aviator. He was at Pearl Harbor when the Japanese attacked and served in both the Pacific and the Atlantic during the war. Promoted to rear admiral in 1957, Moorer served in the office of the Chief of Naval Operations and commanded Carrier Division 6. As a vice admiral he commanded the Seventh Fleet. The only man to command both the Atlantic Fleet and the Pacific Fleet, Moorer became CNO in August 1967. Upon the retirement of General Earle G. Wheeler he moved up as Chairman of the Joint Chiefs of Staff in July 1970, a position he held until retirement four years later. Admiral Moorer died in McLean, Virginia, on February 5, 2004.

17

"YOU'LL NEVER BE A LEADER IF YOU DON'T TELL THE TRUTH"

A conversation with Admiral Thomas H. Moorer, USN
June 3, 1976

Sir, would you please share with me your thoughts about the principles of leadership?

I always approach it on the basis that you must remember, at the very beginning, that it's the people that you're leading that make you look good and not the people who are leading you. That means that you must spend considerable time on these people as individuals and learn their strengths and weakness, because I've learned that there's good in everybody. What happens in ordinary living is that circumstances shape up so that the bad points of individuals are sometimes brought out. You never know about the things that make him really do well. That being the case, that's why you have to know the people. If you're going to get them to listen to you, the first thing you've got to do is always tell the truth. You'll never be a leader if you don't tell the truth. Many a President has gotten in trouble because he didn't abide by that fundamental.

British General Walter Walker liked to tell the story about when they were fighting with the Japanese in India. They were trying to operate behind the lines, so he called in the Gurkha chief and said they were going to have the airplanes go in at about a hundred feet and your men will jump out behind the Japanese. "I'm sure you can handle it very well." The old chief kind of hesitated a minute and the general kept talking. He said, "All right, go down to the parachute loft and draw your parachutes." The chief said, "Oh, you mean with parachutes!" The point was, he was going to do it anyway because they had so much confidence in their commanders that they knew he wouldn't tell them to do anything that couldn't be done. That's the kind of relation that you've got to have with your people.

The next thing you must have is knowledge. In the military sense, that means knowledge about your technology, weapons; knowledge about geography; knowledge about the culture of the potential enemies. So it takes quite a bit of reading. I think that is true in the case of civilian leadership too. I got great satisfaction, though, out of having the opportunity to work with so many people. So far as the military is concerned, you've got to recognize the fact that they're very young services. I think the average age of the Navy is 23 or 24. People don't realize that. So when these youngsters come in at 18, 19, 20, they, of course, represent a cross section of American culture. Some people seem to think that just putting a uniform on them will change them into some person who will behave differently than they would the day before. That's why it's necessary to learn what you can about the relationships between the different races and people from different parts of our own country. It's true that today the technology requires a very different kind of fighting man. At the same time, it's also difficult to get the youngster challenged enough that they will sit down and study and learn what their job really is. For instance, you take a flight deck on an aircraft carrier. We've got these kids in now and in about six months you can weld them into a well-knit team. If you learn how to put the right man in the right spot, then they will get competitive and always want to win.

I found out, particularly when you go to sea and stay any period of time, you must always give the crew something to look forward to. When I was commander of the Seventh Fleet, we'd always tell them we're going to be in Hong Kong in a few weeks or something like that. This would always get their interest. They didn't want to do anything that would force them to stay aboard the ship for some infraction of some kind. That's another tool that you can use with these young people—just give them something to look forward to. It may be going home. It may be going into a port. It may be a contest with a sister ship. Anything to give them a goal—a reason for what they're doing. People don't like to feel that their job is worthless. As a matter of fact, you see so often in the civilian world—civilians don't like what they're doing but they won't quit it because they don't know what else they'd do. To be a good leader is to give the people underneath you a challenge or something to work for.

Frequently you get into the situation of discipline—dealing with those who have broken the rules. There again, you've got to be careful to be even handed and show no favors. At the same time, sometimes with these youngsters it's really interesting to watch them. One time when I was commanding a ship we went into Butler Bay in Okinawa. I called the crew together and I said on the loudspeaker—we knew they had village number one and village number two in Okinawa down from where the sailors went ashore—I said, "The bus service is very erratic, the taxis are very expensive. I don't want anyone to go ashore unless they have enough money to get back to the ship. If you're late, you're going to be held to task for it." Well, sure enough, the next day I had a youngster that came back late. So I asked him, "Now what happened to you? Why are you late?"

He said, "Well, captain, I was in bed with this very pretty girl and before I knew it, it was too late to catch a bus and I didn't have enough money to take a taxi. So I just hitchhiked back to the ship."

I said, "Well, now, I told you not to go ashore without enough money to get back. How did you pay this girl if you didn't have enough money to catch a taxi?"

He said, "Captain, it's my birthday and my boyfriend was treating me." Well, that was too much for me.

I said, "Get the hell out of here."

I've had several things happen like that. One of the things that has stirred me is the code of military justice which was revised at the end of World War II. I think General Doolittle was head of the board. Anyhow, this revision was designed more like the civil court. The facts are that I'd say 95 percent of your infractions are done by just kids. All they need is a good kick in the seat of the pants. But now each side must have a lawyer, they write all this up in his record, it's very expensive to keep the paperwork. Somebody said that the Marine Corps has more lawyers than they have platoon commanders. We have to run the thing now like a civil court. I'll give you an example of what I'm talking about. When the first Enterprise was put into commission in 1938 and I was a junior-grade lieutenant on a fighter squadron aboard that was trained to be part of the air wing on the Enterprise, we had a leading chief that was the salt of the earth. Really. I called him in—his name was Ellington, everybody called him Duke—I said, "Chief, you know, we don't want these young men to spoil their own record. There's no excuse for it and I want you to take their liberty cards"—in those days you had to have a liberty card to go ashore or to go out the base, you had to show your liberty card. It was just about the size of a credit card, had your name and everything. I said, "If you have some minor infractions in this squadron, you have my authority to take their liberty cards and keep them for one or two days"—keep them aboard ship or wherever we happen to be. Sometimes we were on a ship, sometimes we weren't. We went for three years and never had a captain's mast. Never had to write anything derogatory about any of those sailors in their records.

You need what I call horizontal attachment so that people will try to help each other. If you organize it too much like a labor union, then you don't get near as much out of them as if you get them together. A further demonstration of that is on the battlefield when a kid gets shot and wounded and the transfusion they give him is from his buddy next door. Knowing that's going to happen or could happen, you have a bond between the people horizontally, instead of trying to compete with each other and make the other guy look bad so they can look good. That's something you can't put up with in an organization in the military.

Is it possible for a man to be a sham? Are the people that work for him going to know that he's a sham and not a real leader?

Absolutely, they'll pick that up immediately. I think that they're very sharp at evaluating their bosses and knowing how they're going to behave and seeing whether they can be convinced that their leader really has their welfare at heart. Once they find out that he's just trying to make himself look good or trying to deceive his boss by making things look better than they are—well, he's then lost the confidence they have in him. This idea of making them confident that they can rely on you is very, very important. Admiral Halsey was made a two-star admiral right after the Enterprise and the Yorktown were commissioned. They gave Admiral Halsey with his two stars the command of that task force of those two carriers. Admiral Halsey had a chief of staff—this was Browning. We had already finished most of our training and most of the squadrons were ashore in Norfolk. In the afternoon at the drill hall in the receiving station there we had ground school. Browning was the one who came out with the new formations like the circular formations and so on. They were using ship models and moving them around to show what he was talking about. He would always start out, "Now when we have the war with our little brown brothers"—talking about the Japs—"this is what we're going to do." That was '33. After the war started he was, I thought, a brilliant tactician and so on, but then he got command of a ship, which everybody in his age group was dying to get—command of one of the big carriers. He had been so raw with the officers and crew that they had to take him off the ship. I mention that because he was two extremes. On one hand he could probably write a war plan better than anybody, but he didn't have this personal touch. You really have to have both.

Admiral, during your most formative years, what individual most impressed you with his leadership style? Perhaps even a style that you emulated later?

I was fortunate to have some very good commanding officers. I think that certainly later on when I had an opportunity to know an admiral like Admiral Halsey and so on, he was a leader because, in a sense, he definitely understood the need to deal with individuals in his command. He put out an order, Halsey did, and said no one will go to Australia without my permission. I don't know whether Admiral Burke didn't get the new rule or whatever, but he sent a division down to Sydney. Halsey heard about it so he sent for Burke. All of Halsey's staff was cowering around, peeping around, wanting to know just what Halsey did to Burke. That was after Burke had that big battle with the destroyer.

When he became 31 Knot Burke?

Yeah. Everybody was hiding in the corner to see Halsey beat up on him. Halsey said, "Burke, you know I just put out an order that no destroyer should go to Sydney without my permission and you've got a division down there. How do you explain that?" So Burke says to him, "Admiral, my men are out of beer and my horse is out of whiskey." Halsey reared back and laughed and that was the end of the thing. He said, "Get out of here." I mean, those are the kinds of things, the kinds of touches, if you know when to do it, it's important.

I think by and large we had good leadership. There wasn't anyone that I didn't know at the Norfolk Officer's Club. Never did anyone walk into that front door that I didn't know who they were. You knew everybody because we only had 80,000 men then in the whole Navy. One year later, after the war started, we had more officers than enlisted men when the war started. You expected, I think, a different fellowship at that time than we have now, where people tend to ask what's in this for me and behave on that basis. When you've got the media barking away at everybody, it's much more difficult today to run a military organization that it was then. I've said many times, I'm very happy to have lived when I did—on active duty when I did. It was quite different. We had that big World War II with unconditional surrender and we went several years where we weren't worried about budgets. The shortage was materiel not money. Today we've had so many situations where the politicians get into the tactical operations. We've got more field marshals in Washington than Bismarck ever heard of. They all want to get into managing the forces. Practically every one of them thinks that you want to use just as much military force as you hope will do the job, but no more

You take this latest thing when they sent one plane, I think it was, into Bosnia. They had four bombs. Two of the bombs they dropped so low that the fuse didn't arm. One of the bombs hung up and didn't go. One of them exploded something. If it had been me, I would have sent 40 or 50 planes. You get just as much flack from the media and everyone else, the politicians, if you send one or 50. You never could get that through. During Vietnam, we were changing the number of aircraft that we were flying every day up across the DMZ and North

Vietnam. There may be 50 airplanes and the next day 40. "Well," I said, "what's the purpose of this? What are we trying to do?" They sent back and said, get the message to Ho Chi Minh. Well, Ho Chi Minh never got the message because the only way you could give him the message was there's got to be brute strength and a lot of it. Now, in Desert Storm, Bush had enough knowledge at least to let the military people plan the whole operation and fly every day as many sorties as they wanted to. He didn't ever try to fine-tune what targets they would hit. They selected targets and it went off very well. Most of the time, like Clinton has been handling the area around Bosnia, they jump from one position to another. Now we've got Haiti as the number-one objective.

Getting back to leadership, in that context, I think a leader's got to have principles and he's got to have goals. And he's got to stick with them, come what may—no matter how much adverse publicity you get or whatever, you've just got to do what you think is right and when you do that, the people who work for you will pick that up and will bend heaven and earth to help you get it done.

In any group of professional officers you're going to have some good leaders but darned few great leaders. From your point of view, Admiral, what separates good leaders from great leaders?

I think a good leader is one who can successfully handle his job. He was given the job because it was thought that he could handle it in terms of a limited assignment. And he's good at it and he succeeds. That comes down to the most difficult thing that a leader has to do, particularly in wartime—that is to select the leaders or the men that can do the specific jobs that are set forth in the overall plan. Now, the great leader is someone like, I think, General Eisenhower, who had the knack of dealing with a very complex group. He had to deal with Churchill. He had to deal with DeGaulle. He had to deal with Montgomery. He knew how to pick his own generals so they could go over the day-to-day problems while he kept all of them happy. That's a great leader, I think. You can't swap them around or exchange them. You couldn't exchange Halsey and Nimitz. It would have been a disaster. Nimitz was the only one who could deal with MacArthur. I think if you put Patton in Eisenhower's place—a disaster, you see—but yet Patton was so good—he wasn't dealing on a worldwide basis or an international basis. He was just dealing with his own people. That's what I think is the difference. They're not necessarily interchangeable, that's the point I want to make. Now fortunately, we've managed, I think, with our system, to select good leaders. If you look at every war we've been in, even if they start out with bad leaders or unsuccessful leaders, it's not long before they're replaced.

You have had the opportunity to observe a number of Presidents up close and personal. Despite his infirmities, and others say there were many, do you believe that Nixon was a great leader?

Yes, I think that in terms of the broad picture, he sure was. Don't forget he opened up Russia and China. He recognized the part they would play in our future. He thought in terms of international policy far more than most Presidents, I thought. Even after it was clear that he was probably going to resign, the way he handled a National Security Council meeting—as you know, they have a big agenda—he'd be sitting there with kind of a glum look on his face, but when Richard Nixon got to the agenda on foreign policy, he'd sit up in his chair and give you an overview and everything. He could do that. I think that he was kind of an introvert in terms of dealing with people. I'll tell you an experience I had. When we did the Son Tay prison raid, after that, about the time Watergate was at its peak, there was this man writing a book about Son Tay, not about Watergate. He tried to depict the Oval Office when I was briefing him—what we were going to do, how we were going to do it. In this book that this individual sent over for me to comment on he had Nixon cursing in every other breath. You know, four letter words because the tapes had come out. I told him, "You know as many times as I saw Nixon he never cursed in my presence, not once. If you want me to approve this book, you've got to take all of that out of there."

I just got a letter from him right before he died, thanking me for a letter I'd written about Mrs. Nixon, who was a real lady. I'll tell you, she was a lady in the finest sense of the word, had a great sense of humor. I thought she conducted herself like a First Lady should.

People who knew her thought the same thing. He was, of course, very shaken up when she died, I know that. But he was a very interesting man. Particularly coming on the heels of Lyndon Johnson, they were so different. Lyndon Johnson didn't know a toot about foreign policy. That was one of the things that got us in trouble in Vietnam. Little by little he was forced to increase the troop count, so that when Nixon came in, there were 549,500 Americans down there. One week later it was Nixon's war. That's the way the media is. Johnson was so obsessed with his Great Society. He was always after McNamara about the defense budget. He's like Clinton, he wanted to use the money for his great society, I think. He knew a lot about the antics of American politics, far more than Nixon, I think, about how things are done and how you twist arms in the Senate and so on. But he was more involved in the real broad affairs involving foreign policy.

I remember from one of our talks some years ago your telling me that you opposed the appointment of Admiral Bud Zumwalt as your successor. In retrospect, Admiral, how effective a leader will history judge him?

I think that Bud Zumwalt has a lot of mixed attributes. This is not the point but he's something like Clinton. He came in and wanted to change everything. He, of course, relieved me. He was doing things that I thought were wrong. For instance, he had people aboard ship that were authorized to communicate with him directly and bypass some of the people above them. Well, that's just a no-no in a military organization, you know. People seem to often have information that you don't have. Then he really tried to construct the environment that the sailors experienced in a way, I guess, that really did break up discipline. He let them keep civilian clothes aboard ship, then have long hair, more or less. I thought—and he knows this, so I don't mind telling you—I thought that he needed more experience. He'd never had a major command, not once. He had had the Mekong River Command. That was about it. So I thought that he had a lot to learn. He also did a lot of things that I didn't like. Let me put it this way. I think he caused more problems than he solved. That's the bottom line. I look at everybody that way.

General Carl E. Mundy, Jr.

The twenty-ninth Commandant of the Marine Corps, Carl Mundy was born in Atlanta, Georgia. He graduated from Auburn University in 1957 and earned a commission in the Marines. In 1966-1967 Mundy served with the 26th Marines in Vietnam. Later he was aide-de-camp to General Lew Walt, the Assistant Commandant. Promoted to brigadier general in 1982, he subsequently held command of the Fleet Marine Force Atlantic. His term as Commandant began on July 1, 1991, the day he was promoted to four-star general. He retired in June 1995 and resides in Alexandria, Virginia.

18

"PERSUASIVE AND EXEMPLARY LEADERSHIP IS MOST EFFECTIVE"

A conversation with General Carl E. Mundy, Jr., USMC
October 18, 1994

General, are there fewer leaders today who are visible than there were when you first came into the Marine Corps? Is there, indeed, a shortage of truly effective, inspirational leaders?

I don't think so. Today's military leaders are not looked upon with the same elevated awe and esteem that we did on the Eisenhowers and the MacArthurs and the Bradleys and Black Jack Pershing or John Lejeune, but that's probably true in every walk of life today. After all, they were involved in world wars. What gave our leaders then, and really, what gives our leaders today such visibility has had as much to do with the circumstances they found themselves in as anything else. One would have to ask, how much of a national figure—he would have always been a political figure—but how national a figure would Colin Powell have been had not there been Desert Storm? Who would have ever heard of Norm Schwarzkopf? But today he's a national hero. So, events breed that focus and recognition, I think, and while there may be less visibility, I don't think there's a shortage of leaders.

That's an interesting point, General, because I remember General

Shoup telling me once that in his opinion leaders come about as a result of people who need to be led, and that events and situations create the great leaders.

Dave Shoup took command of the 2d Marines on very short notice and, in a very few days, found himself ashore at Tarawa and directed the battle and became an instant personality, awarded the Medal of Honor. I think General Shoup would have been a great leader anyway, but he may never have been a nationally known and recognized leader had he not been involved in that particular battle.

One of the things that has changed a lot, as you know, is that the media has never been more pervasive in American society and to some degree in the military itself. It used to be that, let's say during World War II and perhaps a few years afterwards, many of your senior generals and admirals could have a few warts and get away with it. Not anymore. Does this mean that the leaders today who have to deal with so many other issues are, in fact, better leaders than their counterparts during the Second World War?

Oh, I don't think so. As I recall, I used to see President Harry Truman, a fairly flamboyant figure, actively engage the press. When he would take a walk down Constitution Avenue the press even seemed to genuinely admire him. Today, if President Clinton runs down Constitution Avenue, the press will seem to go out of its way to report that traffic was backed up and was held up, or that he stopped at such and such a place. So it has become fashionable, I think, among some members—not all, we still have responsible members of the press—but we have so many more people who don't appear to have a complete comprehension of what they're writing about and, increasingly, because we have a more youthful America, there are many who have not served in the Armed Forces and who don't understand the Armed Forces and thus write about them in an adversarial sense. They perceive that generals are what generally you would see caricatured in a Saturday morning cartoon.

Having said that, I do believe that the press has an important role to play. There are good and bad reporters just as there are in any profession. I do wish, however, that the media, as an institution, would try to be a bit more balanced in its reporting. There are two sides to every story, and I'm not totally convinced that both get equal space or air time. Here's an example. It is a column written by Colonel Harry Summers. His piece is about an Admiral who gave his military judgment on a case that any one of us would support. And so for that, for giving that advice, he was struck down. There seems to be an effort to find the feet of clay that we all have rather than to recognize that this fellow is doing the very best he can in a very responsible job.

Doesn't that send out a very dangerous signal that we don't want people to stand up and take a position and be honest?

My answer to that would be yes. I believe that what you send with a signal like this is that the best way to be able to retire with honor after you've served 40 years or so, or to be able to keep your family from being embarrassed by having you paraded through the press, is to do nothing. In other words, one could argue that the best way to be able to complete your career with a relatively sound stomach lining is never to be out in front on the important issues of the day. And that's worrisome. Now that doesn't mean we don't do it, because I think we do. We still try and do our dead level best. But we are in an era when military leaders are questioned by the media looking for some ulterior motive. They seem to be asking, "What's his agenda?" "What is it that he or she wants to accomplish?" "Why are they doing this?" The answer, in my case, is pretty simple: I am not seeking to do anything except to run the most effective Marine Corps I can and do it as best I can.

Correct me if I'm wrong, General Mundy, but how can one be an effective leader and not put his or her opinions and beliefs on the line in situations where they are called for? From my perspective, to do anything less negates their leadership.

I believe you're right, but I think you have to look at leadership in a different context. In other words, what we are essentially discussing here is national leadership, i.e., should I, as the nation's senior Marine, be standing up in the press and representing issues that I believe are best for the Armed Forces? I think the answer to that has to be yes. The danger lies in having those views misrepresented by persons who don't understand what the view is, who don't have the experience with which to judge it. Rather, they will judge that view as it pertains to this very unique element of society that we call the Armed Forces, trying to judge it with regard to all society. We are not. We are separate, and we are extraordinary from society. That's not a statement of elitism, but it is a statement of purpose. So when you try and judge us on the basis of what a person running for political office or holding political office might be judged on, then you don't understand the difference between the two. Now the other echelon of leadership, though, and in my view ultimately the most important, is the leadership that we exert within our own organizations. Those who serve below us are not generally prone to be as suspecting or as critical, as doubting, as questioning of the leadership that we give as those who are outside and don't understand why we're doing what we're doing, which ultimately is to influence the organization. That's the difference I think, of being judged on the basis of other standards of measure by an uncomprehending evaluator as opposed to being judged by those who understand the organization.

A lot has been written about style. How would you describe your leadership style?

Well, that probably is one of the hardest questions that I would ever have to answer because I have never consciously tried to establish a style. I am what I am. I am not trying to come across as a modest individual. I would just like to have 175,000 Marines out there generally believe as I do about the Marine Corps. I learned a long time ago that if you try and be something you aren't, then people see through that in about the first five seconds that you're out in front of them. Because of that, I have always believed, I guess, in persuasive leadership rather than in an authoritarian style. There are, of course, times when you have to say "we're going to do it this way" or when you figuratively apply the boot, but those generally are crisis situations, those are times when one must make men stand up against fire when every rational sense tells them they shouldn't stick their head up at that point. But somebody has to get them up. That's combat leadership. But I think on a day-to-day basis, in today's military with the well-educated and generally highly motivated and intelligent people that we're charged to lead, persuasive and exemplary leadership is most effective. I guess that would be my style.

I vividly remember, General, when General Lew Walt visited Albuquerque some 20 years ago as the Assistant Commandant in uniform. I will never forget. He got off the plane and there among the people greeting him was a young staff sergeant and the first person that he greeted was that staff sergeant with a hug. Is such a show of affection between general and sergeant alien to the Marine Corps of the 90's and beyond?

Absolutely not. General Walt was the "squad leader in the sky," the "three-star grunt," and all those sorts of characterizations that derived. He was a great combat leader and he had great compassion for the troops. Now troops can be colonels and troops can be sergeants or privates, but he had great compassion. I think that each of us, as we become older and more senior or longer serving, each of us feels an increasing affection for the people below us. So the affection between senior officers today and between those junior to us grows over time. I think when Eisenhower went down to talk to the 82nd and the 101st Airborne troopers that were getting ready to fly off to D-Day, I think probably his emotion was as strong as mine when I go out and see young Marines that are headed off to the desert or to Somalia. You sometimes choke up with pride because they are so confident and so spunky and so filled with what you always hoped, in my case, that the Corps would be. So there is a tendency, I think, in all of us more senior to have deep affection for the more junior. We might express our affection differently for a major general, that's a friend generally, but to go down to a young 19- or 20-year-old Marine that you know is about to give his or her life or well-being for our country's way of life or liberty or whatever it is, to lay that on the line is emotionally charged. So I think that there are probably as many hugs for staff sergeants in that context as you saw Lew Walt in Albuquerque give. You would not find that to be exceptional today. I'm not talking about me; I'm talking about anybody that you might see. Colin Powell in some context would have hugged that staff sergeant. You would find, I think, most of us feel really, really emotionally attached to the youngsters that are below us.

I would be very surprised, General, if you told me that your association with General Walt didn't leave a lot of impressions on you and didn't shape to some degree what we just talked about, this leadership style. What are the things that you remember most about his leadership style?

Lew Walt was probably, as we all are, a man of two personalities. One was the public personality—the big Teddy Bear, the "three-star grunt", the "squad leader in the sky", the hugger of staff sergeants. The other would be the man that you saw in his office on a day-to-day basis and perhaps had to tell him something he didn't want to hear and, well, let me just say that he did not suffer fools lightly. He was really a hard man. He used to almost take pride in telling his staff, "I'm the toughest general in the Marine Corps to work for." And, indeed, none of us disputed that fact.

Now that I've said that, I have a great deal of affection for him because I was privileged not only to serve him as his aide-de-camp here in the headquarters, but before that, for a period of my service in Vietnam on the III MAF staff, so I briefed him every morning while I was there. I saw there the man that I think you're talking about when we would talk about the casualties among the Vietnamese people. For example, if we had conducted an attack and there had been civilians who had been wounded and killed in that attack, that would almost hurt Lew Walt worse than taking uniformed casualties. He would frequently walk out of those briefings in the morning and get in his helicopter and fly off to the village that had been attacked and spend as much time talking to the village chief about the civilians who had been injured or killed as he would talking to the battalion commander who had lost casualties but had won the battle. So, he was a very compassionate man in that sense. About his leadership style, I would say that Lew Walt would fall more into the authoritarian leadership style than the persuasive. He was a man of action. In other words, his leadership style in peacetime was only slightly modified from the leadership style that he had in combat.

Would you say, General, that his style was pretty much matched by the style of General Al Gray? Were they similar in that respect?

Yes, but they approached problems differently. General Gray was also a man of great intellect. He was a student of the military art, of not only the science of our profession, but the art of war. And as a result, he became a man who has a great understanding of the employment of military forces in the operational sense, in the strategic sense, and in the political-military sense. Lew Walt was more of a tactical commander. He was more engaged in face-to-face shooting at the enemy. Now many would say because of the image that we have of General Gray that he, too, was a guts and glory gun fighter. I did not see that side of him during my service with Colonel Gray when he was a regimental commander, with Major General Gray when he was a division commander, Lieutenant General Gray when he was a force commander, or when he was a Commandant. What I did see in General Gray on those occasions was more of the intellectual, the military intellectual and teacher. He was also very effective with troops. They loved him. They idolized him. They all drank out of camouflaged canteen cups, and they all took up tobacco chewing and all of those sorts of things that emulated their leader. So there was the days when everybody was a warrior, and we all spat tobacco juice and so on because of the Commandant.

I had a senior Army general tell me recently when we were talking about leadership at the division level that in the Army you are going to see increasingly more of that so that your division commanders are going to be recognizable warriors, if we can use that term, who for the most part will retire from that job unless there happens to be one of the few positions at the three-star level that a man of that ilk can move into. And the reason for that is as the Armed Forces downsize, they need to have people with certain characteristics who are able to go out and command at the divi-

sion level and the assistant division commander level, and more often than not those people are not going to be those who are able to then come to Washington and to take on roles at higher levels within the organization because they don't have that combination of skills that is necessary to prosper and thrive here in the Washington punch bowl. Now, is the same thing true in the Marine Corps?

> I would disagree with that thesis. My most effective staff officers are also my most effective commanders. Let me give you just one example. Brigadier General Jim Jones is currently the Chief of Staff of the U.S. Joint Task Force that has conducted operations in Bosnia. He is one of the most respected officers in the European Theater today because he is tremendously effective. And yet, Jim Jones was a guy who put five years in the Senate Liaison Office as a lieutenant colonel over on Capitol Hill. He wore civilian clothes all the time. And one would have said, well, this is a guy who has learned the politics of Washington but who won't, therefore, be an effective field commander. Yet, later this month, he's going down to command the 2nd Marine Division. This after he has been in a significant staff assignment—responsible for negotiating with the Bosnians, the Croats, the Serbs, the UN, and seemingly everyone else in the theater, and he was very successful. He will also be a very successful and dynamic division commander. But he is typical of our current crop of Marine general officers who have contributed significant, high level staff work and held important operational commands.
> The most effective combat leaders that I personally have known have been the most respected and, if you will, able to operate in the political-military environment. My hero is General Bob Barrow, Robert H. Barrow, who was the 27th Commandant. Bob Barrow as a company commander, as the colonel commanding the 9th Marine Regiment in Operation Dewey Canyon, one of the most heralded operations conducted by U.S. forces in Vietnam, and later as the Commandant of the Marine Corps, never changed at all. This is a man of noble character and, again, a man who understands the art of war, who probably has sighted down his pistol sights or maybe his rifle sights at far more people over several wars than anybody else that I know. And

yet, were he at this table today, you would feel that you were in the presence of greatness. He is truly, in my estimation, one of the great leaders of history and is a role model for not only Marines, but for officers of all the Armed Services. So I don't think that you have to be a crusty, old, tough, craggy, snaggle-toothed division commander, hell on wheels, and who can't then come to Washington and be effective. Indeed, I would argue the other way around. The most effective division commanders are sometimes the "smoothest" people around.

So in sum, Brigadier General Jim Jones is not an anomaly in your mind.

No. There are others around. There's Lieutenant General Tony Zinni who is just going out to command our Marine Expeditionary Force in California. He was a brigadier a year and a half ago. He was General Shalikashvilli's chief of staff in Northern Iraq. He was the J-3 for the initial Somalia intervention. He is a recognized leader and is another example of an officer who has served ably and well in both staff and command billets. There are others. So I think that's a miscaricature that you have got to be tough as nails to be a war fighter and that you've got to be a sycophant to be a Washington bureaucrat.

One of the hats that an effective leader has to wear is that of a mentor. Would you agree with that? And if you do, to what degree do you mentor to the people under you?

I guess mentoring, in that sense of being able to detect who in your estimation are the front-runners, so to speak, and to "bring them along" and to coach them along the way, to mentor them, if you will, yes, I think it is a very important function of leadership. There are many in an organization that you can say "This is a very fine officer, and he or she will serve very ably and capably but they're at the limit of their ability within the organization right now in this pyramidal structure that we have." Could they continue in other positions? Yes, and they would be very effective. Do we have room for them? No. So we must winnow those out and bring along, through mentoring in that context, those who will be the senior leaders. Beyond that, I think that you have a responsibility to mentor lieutenant colonels or majors or whoever is around you. Regrettably, there is little time to be able to do that, unless it is done by example, unless they can look at you and say "I'd like to do it that way when it comes my turn to do it."

Isn't one of the most important aspects of your job, General, to see that effective leaders are being developed, those who will step into your shoes some day in the future?

Yes. Exactly. And we do that through much of the system that I was just talking about of trying to sort out those jewels who are definitely going to come along and who have the most to deliver—not just for the Marine Corps, not just for themselves, but for the nation. So, yes, that is a responsibility.

Your watch will straddle two quite different administrations. Realizing that you are still on active duty, I don't want to put you on the spot, but can you tell me, General Mundy, is there a noticeable difference in the effect on the Corps when you go from one type of administration to another. For example, as different as the Bush administration is from the Clinton administration?

Well, again, while I've said we're an extraordinary part of society, at the same time, we are a part of it. Marines read newspapers just like anybody else. And if the newspapers are critical, you can find some people that agree with this or some people that don't agree with that. In the two administration shifts that I have been on hand to witness, and I'm talking about parties now as opposed to maybe changes from President to President with the same political party staying in, but when the Carter Administration came in, it was dramatically different from the Republican administrations it followed. Similarly, President Clinton's administration is different distinctively from the previous Republican administration. The Republicans had more of a military familiarity and orientation and comprehension I think down through the echelons of the administration.

If you stop and think about it, if you looked at the White House under President Bush, you had a retired Air Force lieutenant general as the National Security Advisor. That's no value judgment on Tony Lake who seems to me to be doing a pretty good job over there, and I think a lot of him, and I like him personally. But you had a retired lieutenant general, you had a former Marine as Chief of Staff in the White House, at least initially, and then a Secretary of State who was a Marine, and you had a President who was a Navy pilot. Does that mean that we were then any more warmly received in the White House than we have been by President and Mrs. Clinton, the answer is no, as much as the press would like to make out. The Clintons, in their engagements with the military hierarchy, have been superb hosts.

But, yes, in terms of the comprehension of the military ethos of what it is that makes us different, that's an experience level not present in this administration. I would say to you that when we go into the Cabinet room to sit down with the President, every member of that national security team or the White House executive team that comes in is warm and engaging and appreciative and hand-shaking. So there is no enmity between the senior military and the current administration. I don't think you would find any one of us that would say, "Oh, yeah, they don't want our views" or anything else. If anything, a remarkable difference might be that in President Bush's administration I probably would have been called "Carl" more often than not. At the present time, I'm called General Mundy. But now is that distancing? No. The way it is said is a mark of respect. So I think that we're well respected, but not as well understood.

General Matthew B. Ridgway

Considered by many to be America's greatest fighting general, Matthew Bunker Ridgway was born on March 3, 1895, in Fort Monroe, Virginia. He graduated from West Point in 1917 but missed combat duty in World War I. Shortly after General George C. Marshall became Chief of Staff of the Army, on September 1, 1939, Ridgway—by then a major—joined the War Plans Division of the War Department. His rise in rank during World War II was rapid, culminating in command of the 82nd Airborne Division on D-Day and the XVIII Airborne Corps during the Battle of the Bulge. When Lieutenant General Walton H. Walker was killed in a jeep accident late in December 1950 Ridgway succeeded him as commanding general of the Eighth Army. Then, in April 1951, he earned a fourth star as successor to General MacArthur when he was recalled by President Truman. Following the Korean War, Ridgway served a year as Supreme Allied Commander in Europe and was Chief of Staff of the Army from 1953-1955, when he retired. He died at his home in a suburb of Pittsburgh in 1993.

19

"YOU CAN BE SUCCESSFUL IN PEACE AND A FAILURE IN WAR"

A conversation with General Matthew B. Ridgway, USA
September 24, 1977

Having served in both war and peacetime, General Ridgway, would you say that it is possible to notice marked distinctions between the factors that make a man successful in war or successful in peace?

Yes, you can, very definitely. You can be successful in peace and a failure in war. Yes, indeed you can, and some of the reasons are apparent and some are attributable to the blind spot on the brain of the man who picks them for high command. That's true of almost every high commander. There will be some little blind spot. The only exception, I would say, was George Marshall. I don't think he ever made a mistake in picking one because of misjudgment of the man's character or abilities. The man may have failed after he got there, but not for reasons which had been manifest in his prior service.

Now let me say this about the men who rise to high rank in peacetime. When I came into the Army in 1917, we had only about 162,000 total and 5,000 in the officer corps. That jumped, in World War I, to about 900,000 officers and four million men, and there were still larger proportionate increases in World War II. In peacetime you will find some of these men who got to be general officers had charming personalities, lovely charming wives, they had some outside income—they didn't need to have much, but if you had a couple of hundred dollars more at the end of the month than your colleagues of equal rank you could do a hell of a lot more entertaining. You see, they had the social graces so they pulled their rank. I could name any number of men who reached high rank in peacetime for these and other reasons. Maybe they had political connections or something of that sort. They were either passed over, and didn't get a chance in war, or given a chance and failed.

The controversy over whether leaders are born or made has been brewing for centuries. Is it possible to pick out a young man—a lieutenant, captain, or major—and predict he will ultimately rise to four-star general?

Oh, I would say definitely yes. Take General MacArthur, for instance. I think he was recognized as eminently successful from the time he was in preschool. From West Point days, right on up, he was successful in everything he did. In my own individual case, when I was a captain, one of my senior commanders, a man that I revered—Frank McCoy—said I was division commander caliber right now, and I was only a captain. When I was an instructor at West Point I listened in, one time, on an argument between several officers senior to me. Some had been there for years and years. The question they argued was this: is it possible to pick out members of the Corps of Cadets, in the cadet days, and say this man is going to the top and this man won't? A man in whose judgment I had the most confidence said no. I've been here for 25 years and to say that some of these men at the bottom of the class won't succeed, I would have made so many mistakes that I would say it is impossible to do. Now, I could name two or three four-star generals to you, two in particular, whose cadet records were way down, primarily in discipline. Nothing impinged upon their character, but they were kind of sloppy in deportment, got a lot of demerits, and were thought of as being maverick in a way. Both achieved very high command. It's a very difficult subject. In some cases definitely yes, and in others definitely no. Late bloomers come along. When responsibilities descend upon a man, if he's got it in him he rises to it,

Does this carry through just in the very early years, general, or is there a point later on, say 15 or 20 years into a career, where those indicators of ultimate success are more apparent?

Very definitely. It's so conclusive in my own experience. For instance, I was the first commander of the 82nd Airborne Division when it was first organized, and I had it from 1942 through Normandy. The 82nd had been an infantry division down in Louisiana. General Bradley was my division commander and I was assistant division commander until he left. He told me, "I'll hold you responsible for the training of the three infantry regiments in the 82nd Infantry Division." I knew every second lieutenant, every infantry officer in those three regiments by name. I knew probably 900 of them altogether. I could call them by name, and the reason was I spent every single daylight hour with them in their training. By the time we got into combat I had such a close personal relationship with my regimental and battalion commanders that I concurred in every single one of their selections. Not in one single case was there any disagreement. In other words, they had been watching their men and picked them out. By the time they have a little service it is easier and easier, and the longer the service, the more concrete becomes the contour of their potential, let's say. In Korea, when I got there my first thing to do, the first morning after I'd arrived—I arrived after dark the preceding day—was to go up and visit my division commanders and corps commanders in their own bailiwick. All three corps commanders I had known before, and about half the division commanders I'd served with before.

So, after I'd got the measure of these commanders in their own fields, up in their own terrain, I informed the Army Department—Joe Collins was Chief of Staff—that I needed top-flight regimental and battalion commanders. They sent me a list of the fellows available, so I got top-flight fellows. It's very easy, after 15 years service—and you still may make a mistake—but by that time the weaknesses of any will have shown up, weakness of character, lack of force, lack of the power of decision, the fact that they don't know their people, aren't close to their men. These are the things I can tell when I walk into a combat area. The degree of confidence in the unit, self-assurance, pride in their units—these things just stand out. First time you question any of the men you see it in their faces.

More than two decades have passed since General Douglas MacArthur was relieved by President Truman. Has the passage of time changed your thoughts, your feelings about MacArthur? Has history put this in a different light?

No. You see, I had been quite close to him when he was Superintendent of West Point and I was teaching Spanish and tactics. One day he sent for me and said the athletic department was in trouble, and would I take over the department. So, at the age of 24, I had the status of a full professor at West Point. My only boss was the superintendent, and we had very frequent meetings where I came to recognize the brilliance of his mind. I didn't see him again, except for one brief visit to Washington when he was Chief of Staff, until I reported to him in Manila. I saw him for only a few minutes, for he was off to Tokyo to sign the instrument of surrender. I saw him again in August of 1950 when Mr. Truman sent Averill Harriman, Larry Norstad and me out to Korea when MacArthur was begging for heavy reinforcements. Then the next time I saw him was the day after Christmas in 1950. That's when he had severe reversals there. I asked him if he would approve my going on the offensive, and he said, "Matt, the 8th Army is yours, do what you like with it." Have I changed my opinion of General MacArthur over the years? No, I haven't. I have no reason to change.

I was more concerned, General Ridgway, with the events that surrounded MacArthur's dismissal. You wrote in your book that you were caught up in mixed feelings because you had a great deal of respect for the man, but you also felt President Truman was right and deserved support.

I should have added that while I was Deputy Chief of Staff in Washington before I was sent out to Korea, I deplored General MacArthur's handling of the Eighth Army and said so. It's a matter of official record. In the first place, I was very deeply concerned because he was not carrying out the explicit instructions of the President. I was with Harriman, as I said, and Harriman's objective was to explain in full Truman's decisions and the reasons which compelled him to make them. In spite of that he didn't carry them out. I was in still more disagreement over MacArthur's tactical handling of the Eighth Army. It was deplorable. To give you a specific instance, after the X Corps made contact there through Inchon—which was a brilliant success—the X Corps, with the Eighth Army coming up from the south, should have instantly passed control to Walker, the Eighth Army commander. Any army commander—I would say 999 out of 1,000—would have so ordered command automatically passed to Walker's control. But he didn't. He maintained control from Tokyo, 800 miles away. Then he compounded that by insisting upon evacuating the X Corps out through Inchon when Walker's army was starving for ammunition, food, clothing, everything. Finally, to make matters worse, MacArthur insisted upon keeping control of the X Corps and the Eighth Army, under Walton Walker, from Tokyo—when he didn't really know the conditions there. He then deployed this army over a tremendous front of several hundred miles away when they were absolutely incapable of supporting each other. The terrain was such that the X Corps and the Eighth Army never made contact on the ground except for one small patrol. Later on MacArthur acknowledged this, but at the time he didn't. He said they were self-supporting, but they weren't. My opinion of MacArthur changed to the extent that I thought these were grave mistakes he had made, but it never shook my recognition of MacArthur's brilliance and his wonderful past record.

And that's essentially the same today as it was then?

Yes, it hasn't changed one iota.

General, how did you make it all the way to the top, to four-star general? What contributed most to your success?

Well, the first thing is probably a tremendous element of luck, and that goes for everybody. You've got to be at the right place at the right time, assuming you've got the ability. If you aren't at the right place at the right time, or ill health or an accident or something overtakes you, it can knock your career right down, I don't care how much ability you've got. There are always others who could do just as well as you did in the job given the opportunity. Now then, on the positive side, I would say that my entire ambition from the time I was a cadet was to prepare myself to successfully lead troops in combat. That was what I wanted to do. My concern was for my men from the time I had my first company down in Texas in 1917. Up to 300 men in that company, but in the course of a couple weeks or so I could name every man in the ranks by his name—call him by name. I'd be sitting in my tent at the end of a day's work and there would appear General McCoy. I was just a young captain, you know, and I hardly ever talked to a general officer. Then McCoy was sent to a higher command, and the next fellow was Harold B. Fisk, who had been Pershing's assistant for training in the AEF. A hellion if there ever was one. He took no excuse for anything. If you slipped up you got it on the nose. There was no excuse that could excuse, no explanation that could explain. You didn't do it, you should have done it, that's that.

They were two absolutely opposite types of leaders, but both highly successful. I think I learned as much from General Fisk, of minor tactical matters and the handling of troops, than from any other officer under whom I ever served. He was a tough one. A lot of this goes into the making of someone. You see, if you're privileged to have different types of leaders, and all the time you're watching these people—how did they get there, what are their personalities, how did they succeed, how did they fail?—how much of what they did is suitable for you to adopt, how much is out of character for you? If I had tried to be a showman, I'd have fallen flat on my face. Do I make these things clear?

Yes, sir, you do. In your opinion, what role does personality play in leadership, in how a man deals with others?

It does make a difference. For instance, to give you a specific example, almost anybody will break if the pressure is sufficient. Particularly in combat, extreme privation, extreme fatigue, long periods with no sleep, insufficient food, exposure to fire—all of that. The man may temporarily slip. I've seen some men from our finest units fail in combat because they were simply worn out at the time. If you treat that fellow carefully, get him out, send him back where he can get some rest and food, you'll save him, or you can wreck him right then and there. It's a question of judgment. Is this man really incompetent? Has he any excuse for these things, or doesn't he? I would point to General McCoy, who had the courtesy to listen to people. So far as I could, I would learn some of his background, particularly my first sergeant, my sergeants, my noncommissioned officers.

Second, I would say I strongly disliked the showmanship element. It was not in my nature. I recognized that a man like Patton, like MacArthur, like Montgomery in the British Army—each was a great showman. If that's your nature, that is fine. It certainly produced success for those gentlemen and it has for many others. But that's not part of my nature. I never was a showman and never intended to be. The press tried at one time to play up my wearing a grenade as sort of a symbol, like Patton wearing a pearl-handled revolver. There wasn't any such thing. The grenade is a very powerful defensive weapon if you get into a tight spot. I always wore a live grenade, particularly in Korea. I was out there on the front with my troops in World War II all the time. I've got a piece of German grenade in my shoulder right now.

The third reason is to give everything you've got. You don't worry about yourself and you don't worry about your personal safety, your comfort in combat. You share everything you can with your men. This is what made George Washington an outstanding leader. You realize that your life isn't really any more precious than theirs, and you don't squander men's lives in battle. You don't ask them to do anything that you haven't done yourself and are ready to do again. This is a prominent feature of the airborne service, for instance. When you go out the door of that plane, you aren't any different from the lowest, most recently-joined private in the division. How you deal with men is certainly of fundamental importance. You don't talk down to them. They've got their problems, their lives are just as precious as yours. They're beset by the same fears and anxieties and have the same desires and faults as you have and you should bear that in mind and share things with them. You've got to maintain your dignity, of course. I don't for one moment want to deprecate that. You must preserve your position and your rank, but there are ways of doing that without affecting your authority, your position, and your responsibility. You've got to get down close to the men, that's all. They'll go to hell for you. These are the things that helped me—the basic building blocks in my career. I want to mention one other thing: the privileges of command. While I was the adjutant of the 9th Infantry of the 2nd Division in Texas, in 1927 and 1928, my first brigade commander was Frank McCoy. He was the type of leader that drew out from his subordinates more than they ever thought they had in them. A magnificent type of leadership—always calm, always moderate, always temperate. I never saw him lose his temper.

Let me say, a man for whom I have no respect is Charlie Wilson, the Secretary of Defense. He thought the senior officers who had spent their lives in the military were all a bunch of failures, and he came into the Office of Secretary of Defense with a contempt for the Army. It showed in his indifference. I'd go up to see him after I'd spent hours and hours on some fundamental problem, a major decision which I was required to share with him. He'd just stand there while I was talking and look out the window. He paid no attention to me at all. Well, this goes down deep. One thing I'll never forget is one time Bob Stevens and I had to go up and see him. When we were through he said, "Now you men, you know." Like he was talking to a bunch of labor leaders or something. He was speaking to the Chief of Staff of the Army and the Secretary of the Army. Lack of courtesy and consideration—you don't forget those things.

You mentioned that you, and many other officers like yourself, were very fortunate to have a great deal of luck and good timing influence your career. How much control, realistically speaking, does an individual have on that timing and luck?

Well, that's questionable, but a good point to raise. Without being a self-seeker at all you must abide for those positions which you think, at that particular stage at least, are going to be milestones in your career. It's a recognition of what you have to do to get there. But one thing I fundamentally opposed is bootlicking. I had no toleration for any bootlicking at all. This never went with General Marshall. He would crush anybody and scratch him off his list immediately if he was bootlicking or if General Marshall thought he was seeking promotion. Now, getting a job is a different thing from seeking a promotion. General MacArthur came to me in Korea and said, "I've recommended you for four-star rank." Well, I told General MacArthur that doesn't interest me, I'm here to do a job and until my mission is accomplished I have no concern whatever about my rank. That's the way it is.

General, being appointed to the number-one position in the US Army is an honor only a handful of career officers have known. Can you relate the events that led to your selection as Chief of Staff?

I couldn't tell you. It would be pure conjecture on my part. General Marshall was retired, and whether or not he was consulted I wouldn't have the faintest idea. Bob Stevens was Secretary of the Army and Mr. Wilson was Secretary of Defense, both just appointed. They came over to see me when I was NATO commander, I guess, among other things, to size me up because quite apparently I was under consideration to be Chief of Staff. As a matter of fact, shortly after that I was told that I would have my choice. I could either stay on as the NATO commander or come home and be Chief of Staff. I was 58 years old and my wife and I discussed this thoroughly and decided to come back and take the Chief of Staff job. We made our decision there in Paris that I would retire when I reached 60. I'd been brought up in the Army; I'd never had my roots down anywhere. I couldn't live on my retired pay very well; I had to get a job somewhere, and the sooner I got back and identified somewhere in the United States the easier it would be. I couldn't say at all what the process was in my selection as Chief of Staff.

General Roscoe Robinson, Jr.

The first black four-star general in the Army's history, Roscoe Robinson, Jr., was born on October 11, 1928, in Missouri. Following graduation from West Point in 1951, he saw action in the Korean War as a member of the 31st Infantry. During the Vietnam War he commanded a battalion of the 1st Cavalry Division, earning two Silver Stars. He commanded a brigade of the 82nd Airborne 1972-1973 and returned to command the division in 1976-1978. Appointed a full general in 1982, Robinson served as U.S. Representative to the NATO Military Committee until his retirement in 1985. He died in Washington, DC, on July 22, 1993.

20

"SOLDIERS ARE NOT INTIMIDATED BY GENERALS ANY MORE"

A conversation with General Roscoe Robinson, Jr., USA
October 24, 1987

My company commander, years and years ago, was fairly close to the same class that you were in, plus he had a couple of assignments in SHAPE somewhere. His name is Jerry Gibbs. Did you know him?

> We were in the same company at West Point. He was a class behind me.

I've known him for years, always admired him. Why would a fellow like Colonel Gibbs, who I thought was just a crackerjack, not pin on a first star?

> I think you find a lot like that who never pinned on a first star. I think when I was selected for promotion you had 5,000 colonels in the Army. Now, because there are fewer in the service now—this was 1973—there are fewer. There may be 3,500 or 4,000. And when the list comes out there's going to be maybe 55 or 60 people on it.

That's intense competition.

Right. So who can say why? What I like to say is thank goodness all of those guys who were serving as colonels, and don't get selected, thank goodness many of them are committed enough to stay in and don't feel rejected that they haven't had a successful career. I am of the opinion that you don't have to be a general officer to say you've had a successful career in the military—in spite of your research and all. I would say that if you complete a career and come out as a colonel you've had a successful career, because you've had the good assignments. Thank goodness those guys don't get upset or feel like the system has not been favorable to them and so forth. I had a guy who's a good friend of mine. We were Leavenworth classmates, we were in the same car pool during an assignment when I worked for General Kroesen in Europe. I was his Deputy Chief of Staff for Operations in Heidelburg and my deputy was a colonel, a contemporary, good friend, and just the best person you could know. Why he was never promoted I couldn't say. He certainly was as good as many of the people who were on the promotion list, but after a certain fashion. He had the proper schooling, went to the War College like you're going to find all of these guys are that you're talking to. He did all of those things. He didn't make it but he stayed the course. He stayed 30 years which is the max he could do. And if somebody asked him if he'd like to come back on active duty today as a colonel, he'd jump at it

Is that right? That's really fortunate.

That's right. That's why I say we just have to keep those guys. But all of them aren't going to make it. When I gave you those numbers, I think it's about 3,500 today—I'm not a personnel expert anymore—but from about 3,500 colonels serving on the list will come out about 55 to 60. The hardest rank to make as far as I'm concerned is that one star because when you go for the two star rank it's those serving one stars who are your competitors. So it's a much smaller group who is eligible trying to get up there to the top.

Plus, you've become a bit more visible too. If you do the right job, someone's going to notice.

Yeah. I really think that the jobs that you have probably have the greatest impact on your selection. Which is one reason why, as you're growing up in the services—and the advice that I give to many young officers—is never to shun the difficult jobs. Go for the jobs that are tough. That's how you make your mark. You make your mark showing that you can do the difficult jobs, even though you may not get as high a mark in those jobs. I think when the senior officers look at the records when it comes time for selection—I know when I was serving on boards anyway—the type of job that the person has or had becomes very, very important. And those who have done those difficult jobs and done them well—because just having a difficult job isn't going to get you anything—but having the difficult job and doing it well, you're going to get the proper rewards. Sometimes you get a little discouraged along the way but the system is pretty good I think.

General Robinson, at what point in your career did your personal leadership style begin to gel?

I think certainly as a captain company commander. You probably are developing your style as you go and you do it a lot of ways. You observe others and you see the other "successful commanders" and catalog what this guy is like. I remember working for George Blanchard as a captain and he was everything a commander to me ought to be as a captain. He was a colonel. I would say, boy, if you read field manual whatever and they talked about the commander, that was George Blanchard. As far as how he ran his meetings and those kinds of things you kind of catalog and say, boy, if I ever get to be a colonel, I would sure like to do it like George Blanchard does it. Similarly you'll see a guy who you'll say, geez, he may have been successful there but I could never do it. In fact this guy worked for Blanchard, had a tremendous temper and was more effective the angrier he got, or it appeared to me. I'd say, God, how can that guy do it? I could never flip my lid, so to speak, and maintain the effectiveness that this particular guy did. So that's one of the things. I would try to keep my cool, keep my temper, and not go flying off. I'm not a person that lights into my subordinates. I think that all of those kinds of things have had an impact and I would say that, at the captain level, you probably start looking at those because you're much more aware of your seniors at that point in time. At least for me that was the case.

How different was that leadership style 25 or more years later when you were wearing four stars in Europe?

Well, again, it's duty related. There were a lot of things. What I developed over the years served me well as a senior officer just as it did growing up as a company commander, a battalion commander, a brigade commander. Other than the scope of your responsibilities—and I'm talking about taking care of your people and being concerned about what's going on—I've always been a believer in the chain of command and delegating authority. I say delegating authority, but still keeping your finger on what's going on. The last thing you should do is not be aware of what's going on. Again, it depends a lot upon your subordinates. I think you learn very early who you can trust, what jobs you can give them, and you act accordingly—whether you are at the company level or whether you're at the division level. I can recall one assignment that I had where it just didn't seem to happen unless I initiated it.

Then, on the other hand, I went to command the 82nd Airborne Division and I would think of something that I might want to do and I would call the chief of staff in. This is shortly after I arrived. I'd say, "Bobby, I think that we ought to do such and such." And he'd say, "Yes, sir, I thought that you might be inclined toward that so I've had the G3 draw up a little plan. He'll have it in to you tomorrow." Well, obviously I don't have to give as detailed instructions as I did in the previous assignment because I've got the guys here that are certainly thinking on the same wavelength as I am. That kind of follows. You learn that. I think it's very important over the years to learn who you can give a job to and make sure it's going to get done. Some people you may have to give a little more personal attention. I used to like to tell my company commanders—when I was a brigade commander certainly and probably as a division commander, too—I always like to go down a couple of levels with a very personal interface. When I was a brigade commander I kept a good interface with my captain company commanders and certainly as a division commander I knew my battalion commanders very well. I used to tell my company commanders that they had the authority because they were always worried about somebody over-telling them how to do things. I'd say, "You guys can do anything you want to do. Just remember that. You can do anything you want to do, providing that the battalion commander concurs with it." What I was really trying to tell them was that he's got the responsibility. You're talking about a lieutenant colonel as opposed to a captain. He's got more seasoning, he's more mature, he's done all of those things that those guys have done or are trying to do. Without taking away any of their initiatives let them go on and do it, but make sure that he agrees with what they're doing because he may say, it doesn't have a chance of working. You might get somebody hurt severely—you're going to lose something—and he should step in if he sees something. I'm talking about in the training environment, certainly in a combat environment. He's going to step in right away. They used to do a double-take on that when I'd tell them that but I think they understood what I was saying. Yeah, you guys have got plenty of opportunity to use your initiative, but your senior officer—if he sees anything that's going to get screwed up—it's part of his job description to make sure you stop it and get it squared away and have a little counseling session with you right on the line and go on about your business. I think they understood that. I had some company commanders who've done extremely well.

That's an interesting philosophy. You spent a lot of time working for Fritz Kroesen. How would you describe General Kroesen's leadership style?

Kroesen's a great leader. I had the very fortunate opportunity to work for Kroesen behind Blanchard on two occasions. They were both my division commander when I was a brigade commander and then they were commander-in-chief of USAREUR. I went over to work for Blanchard and when I left Kroesen came in. If I patterned my style off of two people, it would probably be those two. They're different, they're quite different but very effective. I remember when General Blanchard left the 82nd. The brigade commanders, we all had a great, great amount of respect and admiration for him. Without knowing who was coming in to replace him—I didn't know General Kroesen, most of us didn't—we said, oh, what a hard time this guy's going to have coming in to replace George Blanchard. Well, General Kroesen came in and the division got better. As far as we were all concerned, the division just got better. I learned something from that. You go into a job, regardless of who was there before, it's now yours. You don't worry about what the other guy did. Kroesen did not operate the same way. I don't want to say he's not as outgoing, because he's a very personable person, but he just operated a little bit differently from Blanchard but very effective. Doesn't say as much, but when he says it, it's very profound. Those two guys were really excellent mentors or role models or whatever you call it for their colonels.

How was it possible, General, that one man can have that much influence on an organization or a unit?

The 82nd—and I'll use the 82nd as an example—is difficult to put your stamp on, so to speak. I think that first of all, you've got to look at the organization. If it is a unit that believes in banding together, which most of our good units are, the commander is going to have some influence and he's going to gain acceptance by his leadership. People recognize a person who is a good leader, whether it's by his previous reputation or seeing him operate in the field. So, I do think that most of the changes are not going to be profound. They just may be changes in operating procedures, individual changes. They may not have too much impact down the line except that people perceive that we've got a new boss now, this is the way he wants to do it and we're going to get with the program and we're going to do it that way.

You make it sound so easy but I know it's not.

It's not easy but I've been very fortunate, I think, in the units I served with. I served several tours in the 82nd. In Vietnam I served with the 1st Cav, the same kind of unit, very 'can do' attitudes, outstanding noncommissioned officers, junior officers who think they're the best, and all those things usually turn out a pretty good unit.

Did you ever have a chance to serve with or know Mel Zais?

Yeah, I knew General Zais very well. He's an airborne guy. I never served directly for him but knew him very well just because he was an airborne person. Excellent reputation. The NCOs usually let you know who the good ones are and the NCOs would speak very highly of General Zais.

Looking back on your career, General, now that you have the opportunity for a little bit of hindsight, to what do you attribute your success?

I think, first, the willingness to work hard. I think that's a definite requirement if you want to succeed in our Army. You've got to have a willingness to work hard. You've got to have a personal commitment to your soldiers. You have to have that aspect of caring leadership that is evident. You know that old thing about fooling—you don't fool the troops. They recognize a guy who's a phony very readily. I think you've got to have that aspect of caring that is visible. They've got to know that you're willing to share their hardship with them and they're going to go out for you all the time. I think that has probably been very satisfying to me. I have not had organizations where people were not willing to give their all and put out 100 percent. That's what makes a good organization. I think those aspects—first, willing to work as hard as necessary and maybe a little harder and showing a caring leadership style.

Do you recall what you were doing when you were notified of your selection for four stars?

Yeah. The selection was kind of anticlimactic. I knew I was being considered. Yes, I do recall very well. I was in Japan. In fact, because of the time change, it was at night. I was at a social affair and got a call that the Chief of Staff wanted to talk to me on the telephone—General Meyer. He said, "I'm thinking about nominating you to be our NATO representative in Brussels and I'd like to hear your thoughts on it before I move the nomination forward." I told General Meyer if that's what he wanted me to do then I'm ready to do it. And that's how I was notified that I was being considered. Of course then it had to go through the process and so forth. That part was kind of hectic as most of them are because I was supposed to be in Brussels right now and I can't move until I'm confirmed by the Senate and that kind of thing.

You mentioned, General, that you knew you were being considered. Is that a formalized process or is that more of somebody telling you informally that you're one of those that are being considered for a job? How does that happen?

No, this was a case where my consideration meant that I was to be the Army's nominee for that particular job and General Meyer wanted to know that I wanted it. Why waste a nominee if I was not interested in doing the job? If you get up to that level, a lot of people would say, no, I'm not interested in that job. I was sitting in Japan commanding Army troops out in Japan without any thought that I might be moving, getting promoted, or whatever.

So, it was, in all respects, sort of a bolt out of the blue.

Oh, yes, yes—as all of mine have been. I remember all of mine very vividly. Yes. My first one was a bolt out of the blue. That was probably more interesting than the others when I received my first star. I was commanding a brigade in the 82nd when my sergeant major came in the office. You know how sergeant major's get information. He said, "Sir, have you heard that the brigadier general's list is coming out today?" And I said, "No, I sure haven't." He said, "Well, I just got some indication that you might be on there." I said, "Sergeant major, get out of here." So he did. And about a half hour later I got a telephone call from a friend of mine and he said—this is now about 8 o'clock—"The BG list is going to be released at 11 o'clock and you're on there." I said, "That is not something to pull a person's leg on." He said, "No, I'm telling you, you're on there." I said, "Thanks very much," and hung up. I still didn't believe it. About 10 minutes later I got another call telling me that I was on and gave me my number. He said, "You're number five on the promotion list that will be released at 11 o'clock."

I then called for my jeep and went out in the field and I was in the field when I was notified officially at 11 o'clock. You mentioned General Zais. He happened to have been the president of that board. He was visiting Ft. Bragg at the time and he came to the field—he came to my unit in the field and he congratulated me at that time.

Did you have a relatively sleepless night that night?

No, I can't say that I did.

Of course, it's difficult for somebody on the outside to fully comprehend what that means, because that's an awfully big step.

It's a big step, yes. We were very fortunate. There were five of us in the 82nd who were selected at the same time.

And where did they all go?

Two of us made four stars, one made three, one made two, and one retired as a BG.

Who was the other four-star?

Volney Warner.

Wow, somebody was right on in picking these people.

I would like to hope to say that is true, yes.

One of General LeMay's aides, a fellow by the name of P.K. Carlton, tells a story that when he first went in as aide to LeMay, LeMay sat him down and said, "Look, if there's anything I want you to learn from this job, it is I want you to keep your eyes and ears open to learn what makes me tick, and I want you to pick up all of the positive aspects of my leadership style, because when you're sitting in the chair, I want you to do the same thing for your aide."

Did you have the same sort of situation in your career?

Well, I was never an aide, but I was an exec to a three-star general, General Charlie Corcoran, which was a great, great learning experience for me. When I was selected for brigadier general, General Corcoran sent me a letter congratulating me and I wrote him back and I said, "Of all of the things, of all the assignments that I had I think sitting outside of your office, having the close interface—many times a day an exec would be back and forth—and seeing how you operated probably was as good a preparation for my duties as a general officer as any I've had outside of pure field-type things." So being his exec really was a good preparation for some of my responsibilities a little bit later on. General Corcoran retired as a three star, but also had considerable impact on the way I ran things in the office because I saw how successful he was at it. Now as far as the aides that I had, I rarely kept an aide for more than one year because I felt that they should be out doing more important things. But at the same time, I did look upon the assignment as a great learning experience for them and tried to make sure I shared with them enough for them to understand why certain things were going on because the person down the line was not able to get that perspective. So I did try to bring the aide under my wing and make sure that he understood why certain things were going on and why I was taking certain actions, often after the fact, but I had a close relationship with all of my aides, I think. Kind of the same way that you're talking about LeMay. LeMay obviously did a good job with General Carlton.

That, then, is probably the greatest legacy that senior generals leave in the service is the people who they've had work for them who will then be the future leaders.

I think so. All of us share a great amount of pride when our young charges that we had working for us, we see them moving up the ladder. A number of my battalion commanders—when I was commanding the 82nd—are general officers now. When I was at DCSOPS, a colonel and I were talking—he was my deputy, he and I are good friends—about the bunch of colonels who worked for us and there are about five of them who are general officers now. So, that's a pretty good feeling and you certainly hope that when you have these guys working for you that some of the things that you did may have rubbed off on them a little bit and you certainly don't try to make them in your mold, but just from their understanding the way you did things. Whether they would want to do them the same way or not, at least have them understand why you did them. I'm very pleased that some of my subordinates, who were battalion commanders while I was commanding a division, are now division commanders. It's really a satisfying feeling. But I think I had the same feeling with some of my enlisted troops, too.

I can recall one sergeant major very well who was a squad leader of mine when I was a company commander. When I was a brigade commander he was a first sergeant in my brigade, and when I was a division commander he was a sergeant major. A lot of times I would walk in as a division commander and walk in a mess hall, say, a dining facility now, but I'd walk in the mess hall and the mess sergeant would report to me and say, "Sir, you don't remember me but I was in your company when you were Captain Robinson in the 504th." That's a good feeling anytime you see those guys coming up. My company that I commanded in the early '60s at Ft. Bragg in the 82nd I probably had at least a half a dozen of my soldiers that went on and got commissioned during the Vietnam War and I ran into a number of them as captains who were spec-fours the last time I had seen them. So those all give you good feelings.

I can well imagine. Your classmate, General Bill Richardson, feels that the current group—and this was only a year, year and a half ago—the current group of Army leaders, the general officers particularly at the very senior level, is the finest group of this century if not in the Army's history. Would you concur?

I would say that they're pretty fine people, from professional qualifications, personal traits, the whole works, I would say, yeah. You always leave yourself open for controversy when you say that this person is better—especially when you're comparing eras. It's kind of difficult to say but I would guess as a group, if you leave it as a group, I would probably say yes.

As I recall, General, one of the overriding reasons that Bill Richardson used was visibility and accountability—being known by all the troops.

See, that didn't happen in the old Army. Today's general is right down with the soldiers all the time, is seen by them. There were some exceptions, of course. I'm sure all the soldiers knew General Gavin. But that's very true.

Is that healthy to be that close to the troops?

Now, there's a difference between being close and being visible.

I mean by proximity. Being that visible so that they can see what you look like and what makes you tick.

Tremendous advantages, tremendous advantages in my view. And that's what I meant when I said that they've got to know that you're there with them and that you're sharing in their hardships and so forth. I'm talking about it as a division commander with the soldiers. I'm not talking about a four-star general who would be down with the troops once in a while. I'm not talking about that kind. I'm talking about that division commander. That's very important.

Isn't there a lot of risk when they can see you warts and all and they know how you think and what you're going to do? Isn't there a great element of risk?

I don't think anybody worries about that kind of thing. I don't think that anybody worries that somebody's going to see that you're not perfect. We're not perfect. But the guys want to see who you are. You start off together at PT in the morning. At Ft. Bragg in the morning our headquarters boulevard is the main street through the division area. Soldiers are out running, officers are out running, the general is out running. It's very healthy for the soldiers to see their leaders out there. I think maybe that's what Richardson's talking about. Soldiers are not intimidated by generals any more. They will talk to a general because he is visible to them. They see him as a person. I hate to keep going back to the 82nd, but that's where most of my troop experience was. You're out jumping with them and they see you in every environment and they like that.

That doesn't adversely affect the chain of command?

Oh, no.

Or their willingness to say yes, sir, and hop to it?

Absolutely not. I think that's some of the things that military people—when they go into other things—take with them. There is that interpersonal relationship that just doesn't change because you shift your environment, the way you operate with people, the way you treat people, the way that you look upon the support in that superior relationship. Very transferable, I would say.

Were you ready to retire? Would you like to be back on active duty?

We retire when we get a certain number of years of service. So when I came up on 35—35 is the magic number—so that's it regardless of what your desires were. I didn't feel as though I was ready to retire. I was still in good physical condition, still enjoying what I was doing, but it's a policy that I think all of us agree with, so it's certainly accepted. It doesn't come with any disappointment that you retire, it's time.

General Kroesen told me about an article about you that was in the Wall Street Journal. I haven't seen it.

It wasn't really about me. I'm getting a lot of credit for it, but it was about circumstances of senior black generals when they retire not getting positions in the private sector that we feel are commensurate with what we demonstrated we could do in the military. The Wall Street Journal kind of highlighted that—that there's a great resource that "corporate America" is not taking advantage of with the experience that a number of these officers have. Now, I think it extends beyond just black generals. I think it extends to our general officer population as a whole. Unfortunately, it affects most of the senior black generals and some of the white generals. There are many generals. I think there's a bias in the corporate world about generals. There are certain stereotypes, views, that folks have. Well, I can give you a couple of examples. "Well, General, we can't pretty much bring you on board here. We can't give you the kind of support that you're used to having—aides doing all of your things for you and all that." There are a lot of people that are very surprised to find that when a general retires he can make his own airline reservations and get his luggage together and check in at an airport, and on the other end, get a rental car and drive to where he's going. I'm serious.

It's amazing.

Yeah. I'm serious. What I told many of them, I said, "You're thinking about times past. Most of those bennies that you think about—that our senior generals have—have long since been taken away from them." That's what I mean when I say there's some stereotypes out there that really need to be wiped out. I've tried to do that with certain people that I have interviewed with to kind of let them know what being a general is all about, because they really don't know.

General David M. Shoup

David Monroe Shoup was born in Battle Ground, Indiana, on December 30, 1904. Following graduation from De Paul University in 1926, he was commissioned in the Marine Corps and saw service at sea as well as abroad, including two tours of duty in China. As a colonel, Shoup commanded a marine force that assaulted Betio Island during the Tarawa campaign in November 20-22, 1943, and was awarded the Medal of Honor. A decade later he earned a first star as fiscal director of the Marine Corps and in May 1956 became inspector general of recruit training in the aftermath of the drowning deaths of six recruits at Parris Island, South Carolina. As a major general he commanded the 3rd Marine Division and in the summer of 1959 President Eisenhower named him the 21st commandant. Shoup retired in December 1963 and spent his last years in Arlington, Virginia, where he died on January 13, 1983.

21

"LET 'EM LIE"

A conversation with
General David M. Shoup, USMC
May 2, 1973

General Shoup, how did you first get started in the Marine Corps?

I was a student at De Paul University where military training and ROTC was a requirement for two years. Then you could ask to take two more years and during the third and fourth years of ROTC you got paid 30 cents a day. The reason I took the third and fourth years of ROTC was simply an economic matter. The 30 cents a day, $9 a month, paid my room rent. Otherwise, I would have never done it. Then, of course, you're subject to the Army's hocus-pocus of going to camps and all of that kind of stuff. After graduation I was to go down to Camp Knox on active duty for two weeks. At the end of two weeks they kept continuing it two more weeks if I wanted to go, which I did. In the interim, just before college graduation, a representative from our class went to the national Scabbard and Blade, which was an honorary military fraternity, meeting in New Orleans. The principle speaker there was John A. Lejeune, major general and Commandant of the Marine Corps. He gave the principle address and he said, "If any of you young gentlemen have friends in your classes who are honor students that want to get in the Marine Corps, tell them to write me a personal letter." Well, I took him at his word and went back to my room and I sat down and grabbed a piece of paper and wrote a personal letter to General Lejeune, head of the Marine Corps, and told him I'd like to get in the Marine Corps.

While I was at camp, I got a telegram, which I still have. It said for me to go to Chicago and take my diploma, my degree and my birth certificate for consideration of a commission in the Marine Corps at our expense. I went to my Army major and he said, "Well, you're not supposed to get any time off when you're on these two week duties but I kind of envy you and if you go and leave a request for three days leave here, and if you come back as you think you will, I'll just tear it up." I went to Chicago and met my father in Crawfordsville and got my degree and birth certificate and came up to Fort Knox; Camp Knox I guess it was then. I finished that and another two weeks active duty. In the meantime, I finally got some little word from Marine Corps headquarters, signed by General Lejeune, wanting me to fill out this or fill out that, which I did. It was always will you please do this and will you please do that and I did it. One day I got a letter that said I had to take an oath. We went to the little hometown bank and I took this oath and I didn't pay much attention to what it was.

Then some time later, I got this letter that says you will report to Philadelphia Navy Yard. I told my mother, "This doesn't say whether I want to or not, it says you will." So I said, "Maybe I'd better see what this is all about." I was given enough time to drive from Indiana to Philadelphia in a T-model Ford and I reported to Philadelphia. In the meantime, I had a commission in the Army reserve and I had a commission in the Marine Corps and I didn't know it. I didn't know they overlapped. The day before I went to Chicago I was supposed to take an examination in calculus for a regular commission in the Army, but I went to Chicago instead and didn't take that exam and was commissioned in the Marine Corps. Then, of course, I wrote to the Army and resigned my commission in the Army reserve.

Up to this point, until I got to Philadelphia, I had been commissioned, I believe, since May and this was in September sometime. I never saw a Marine. I never saw a Marine uniform. It was many months later before I had a Marine uniform. I went up to a camp in New Hampshire to play football and to Parris Island, but I'd never seen a Marine uniform. I didn't have one. So that's how I now have in my possession a letter which acknowledges my resignation and I have an honorable discharge from the Army. Another little facet to this that's quite interesting. A few years later, to my pleasure, I found out that the statute of limitations had run out but that drawing a salary from two different jobs in the federal government at the same time was a criminal offense subject to five years in prison and a $5,000 fine. I read this when I was aboard a battleship in 1920. I was in Maryland and read this in the regulations. As I say, the statute, I found out from the lawyer on board that the statute of limitations had run out so I was not even compelled to send back the duplicate money.

Looking back on your long career in the Corps, who influenced you the most?

I think without a doubt General Vandergrift, General Erskine, and General Thomas had the greatest effect on me. I worked for them, of course. General Erskine was a Leavenworth-trained officer. General Vandergrift was in Peking as a commander part of the time, a colonel. I made statements to my wife several times when we were out there that he was the only colonel that I'd ever seen that acted like I thought a colonel ought to act and displayed intelligence that I thought that an officer of the rank of colonel ought to in the Marine Corps. I said many times to my wife that he would be the Commandant someday. Erskine was a hell-for-leather fighting man in all respects. General Thomas, perhaps, most scholarly. I think those three people influenced me to a greater degree than any others, while there are many others that I served with. Some influenced me negatively, but if I had to pick three people, I would say that they influenced my thinking and my desire and ambition to perform more than any other people

General, I've been told you had a reputation in the Marine Corps as a blunt, confrontational officer who really put the fear of God in your men. Is that true?

I was not one to pull my punches. What I believed I said. I didn't take any bull from anybody. If I believed it I said okay, prove me wrong and I will disbelieve it. But as long as I've spent this many years and this many hours thinking and contemplating this subject I have come to some definite conclusions. Until somebody picks me up and shakes me and shakes those conclusions out of me I'm going to stick with them.

Did you ever have any serious discipline problems you were forced to deal with?

No, I never did. Never. I've relieved officers, even of two-star rank, threw them out of their jobs. It wasn't for insubordination, it was for poor judgment. In my day I never disciplined anybody for poor judgment, I just fixed him so his judgment wouldn't be a bad influence in the future. The question of how to discipline somebody—I remember one lad, he was the only kid that really played like he wanted to be incorrigible. He had done something, which in those days and in my ethics and code of conduct and moral training, was a rather bad thing, in which he had gotten some unmarried girl pregnant in a little town away from Camp Elliott. He had lied about it and he then became AWOL and while I felt that he had a good background, he refused to carry out some orders. After thinking it over a long time and reading some of our history, I gave him five days bread and water, which used to be the way to fatten anybody. If you give them bread and water, they always come out weighing more than they did going in, because there was no limit to the bread and water. But I'd never experienced that kind of treatment and I wasn't so sure whether it was good or bad, but I was at my wit's end. When I had him back, I asked him if he now wanted to do it and he said, "No, sir." I said, "I'll have to give you five days more bread and water." That went on for five times. This got pretty serious in my mind. I consulted the doctors and I went to my commanding officer and I said, "I think this fellow is still just stubborn and there's nothing wrong with him and he's just stubborn." Anyway, after the fifth time I brought him in and I said, "Do you still refuse to obey this order?" And he said, "No, sir, I'll do it." After he lost, like a good politician, he became one of the finest Marines that we had and later had a terrific record. I used to get letters from him saying he didn't know what he'd be today if I hadn't been just a little more stubborn than he was.

General, you have faced death many times on the battlefield. Did you ever fear for your life?

I believe that even from the time that you're beyond the reach of the enemy and naval gunfire, and I'm talking about the commander's position and it also includes the position of every individual marine and soldier, at this hour you're going in for the enemy, you've got a task to perform and there is a realization that someone's going to get hurt, hopefully the enemy more than you. Now, as you get closer and get tangled up in this business of combat you get into areas in which squad leaders are lost, platoon commanders are lost. There's a little sensing of disorganization, the lack of leaders. When that comes to pass the squad that's looking for a leader, someone to tell them what to do, and there is no one to tell them what to do, they're dead.

   Here on the beach, amongst other bodies and everything, is a potential for causing the enemy greater grief but it's not available because leaders have been killed, disorganization has come to pass. At that point in time the germs of fear begin to multiply. The interesting thing about it is this situation that I described does not come to pass for the commanding officer because the commanding officer is still confronted with 150 things that he must do, that he has to do and that he's thinking about doing which pushes aside all thoughts of anything else and thus eliminates the idea of fear. I believe I'm expressing what every commander of a combat unit would say—I know it's absolutely true in my case. We knew you might get killed if one of those bullets went right through you or a shell landed on top of you but no thought was given to that because you had a mission and as long as that didn't happen you were going to continue with your mission, your job. When you lack that, when there's nothing to do, then is when fear creeps in. But I think, generally speaking, fear never creeps in at the level of the top commander. I don't think it ever does unless you're routed, which Marines never propose to let happen. Other than that I don't believe that fear creeps in except in situations where leadership is lost.

   Here's a young lieutenant or young corporal. The platoon leader or the squad leader gets killed. Here's all these men, more or less in a pile so to speak, and they don't know what the hell to do. There's no one to tell them. They're prepared, they're ready, but they don't know what to do to help attain the mission. That's when fear creeps in, to the extent that along the beach line I went along this place that had some four feet of protection from the sea wall bank and here was a marine, so help me, who had dug himself into that bank until only his feet were sticking out. Now that's how much fear he had. I took my shovel, which was my basic weapon, and I played hot foot with his feet. I really beat hell out of his feet and drug him out of there. In order to shock him I said, "Do you have a mother?"

"Yes sir, yes sir, yes sir." He was in a state of shock.

I said, "I'll bet she would be really proud of you now." I was trying to bring him back to reality.

And he said, "What can I do?" That's all he wanted to know; he was lost.

So I told him, "Get yourself a squad."

He said, "Sir, how do I do it?"

I said, "Just the way I got you. When you get your squad I'm going to be right at that log house and you put the squad in some protective area and report to me."

What rank was he, General Shoup?

Corporal. I forgot about it. Some minutes later, maybe 10, 20, 30—I don't remember—crawling up to this log house was this marine and so help me it was the kid I dug out of the bank. He said, "Sir, I've got my squad." Now that's the man I drug out of the bank who was so completely overcome with fear that he didn't know what to do. So I gave him two or three missions that a squad could well handle and he said, "Aye, aye, sir" and took off back over the hump and got his squad and did some miraculous military action. I later recommended him for the Navy Cross because of the tremendous job he did. That's what I'm trying to say, that if leadership is lost, if you don't know who to turn to and ask what do I do now, or someone to say come on, let's take that pillbox. When that's gone all is gone.

What did you look for in a man when you were selecting him for a leadership position?

That question has been asked so many times—what makes a good leader? I've done a great deal of thinking about this and I think perhaps the answer is not the leader as himself, as an individual, but the attitude of those whom he has been assigned to lead. If they have explicit faith in your ability, in your tactical knowledge, in you—that you will use this little organization or big organization to the specific benefit of the whole rather than some stupid commitment, for example, that tends to make them follow you. So, what I'm trying to say is that the great leader simply comes into existence because of the willingness of people to follow him.

He has to be believable then.

Right. And they have to have faith in his ability, faith in his knowledge, faith in his belief in himself, I guess, and faith in his troops and their ability. It's a matter of mutual faith.

General, what emotions did you feel in combat when you were forced to make decisions that were certain to cost men's lives?

I think the basic decision is made when you're directed to land against a fortified beach known to be inhabited by fanatical and well-fortified enemy troops. I don't think you can isolate your thinking and say, "Oh my God, we're going to get somebody killed." You know you're going to get somebody killed and hurt. That's part of combat. You accept that as part of combat. You must not let it bother you. You must not let these things deter you from your goal of accomplishing the mission. It's hard to do but you express this to the troops: "Dead men, let 'em lie. Badly wounded men, let 'em lie." This is terrible but that's what you've got to do, because that wounded man is of no benefit in accomplishing the mission—he's done his part. Let him lie. Somebody coming along behind him will tie up his wounds and maybe save his life, but you cannot stop to do those things.

I'll give you an example. Going along the pier there was a young man who was crawling along when splat, a bullet hit him right square in the side of the cheek and just blew off the side of his face. I must have been within five or six yards of this. Obviously a lad who was this young marine's lieutenant, his platoon leader, was nearby and he rushed up. I don't know why it is but so often people yell for water. This wounded marine was yelling for water with what facial help he had. This lieutenant, I suppose, recognized him as one of his boys. It's pretty tough to say let him go, let him lie. You want to go help him. So, the lieutenant wasn't sufficiently trained to let that marine go and go ahead with his job with the rest of the platoon. He hesitated and went up to where this kid was and I can still see him reaching for his canteen to give this poor marine a drink that he was yelling for. Just about the time he got up there with the canteen the Japanese shot him right in the head. We lost a platoon leader because the platoon leader didn't let the wounded man lie. That's what I'm talking about. The lieutenant should have been on down the beach urging and directing the rest of his men, but instead he went to help the wounded.

I've got to believe that today's leaders would find the action you advocate quite heartless.

If the wounded man is completely out of the area of enemy fire that's something else. That's not what I'm talking about. The Japs are looking at you right out here. Maybe it is heartless. Combat is heartless. You can't afford to lose a lieutenant while he's helping a wounded man by what amounts to a foolish action. You don't win battles with a wounded man, you win it with men who are still fit.

How is it that a combat veteran, a Medal of Honor winner, came to oppose the Vietnam War?

First, I was disappointed that we weren't going out there to win the war. Secondly, I stated that the methods that we were adopting in South Vietnam to defeat the enemy made it impossible for us to win the war. That was my basic position and I stuck with it. Most people have never discerned the mental machinations that went on for me to come to that conclusion, but to me it was almost a simple, childish, juvenile level of logic. For example, my feelings have always been that in order to win a war you must defeat the armed forces of the enemy or you must make the government of the enemy believe they're going to be defeated. Only then can you win a war. As soon as I found this to be an absolute fact, which I confirmed with several of my friends in high government positions, that we were not—our administration was not going to permit us to go into North Vietnam—I then knew we could not win the war. We could not defeat the armed forces of our enemy because they were in North Vietnam and we were not going to go in there. QED, therefore, we could not defeat them, which conclusion was later substantiated by what has happened. There's an interesting psychology to this. Every marine and every soldier and every sailor is trained—physically, mentally, psychologically—to win. Now here is a fellow with 35 or 40 years of experience in military service telling the world that the methods adopted by our armed forces will prevent us from winning. It's obvious that a great many people were going to differ with me, but I wonder what they think today.

General Maxwell D. Taylor

A member of the West Point class of 1922, Taylor rose to prominence in World War II, first as chief of staff of the 82nd Airborne then commanding general of the 101st Airborne Division at the Normandy invasion. He was superintendent of the U.S. Military Academy from 1945 to 1949, commanding general of the Eighth Army in the closing months of the Korean War, and Chief of Staff of the Army from 1955 to 1959, when he retired. Called back to active duty by President John F. Kennedy in 1961 to be his military representative, Taylor became Chairman of the Joint Chiefs of Staff in 1962. He retired a second time in 1964 to accept appointment as ambassador to South Vietnam in July 1964. The following year Taylor was succeeded by Henry Cabot Lodge. In retirement he published two books. General Taylor died in Washington, DC, on April 19, 1987.

22

"LUCK GOES ALL THROUGH THIS BUSINESS"

A conversation with General Maxwell D. Taylor, USA
October 21, 1974

General Taylor, what are the characteristics of a great leader?

When I was in my days at West Point, in the days when I instructed cadets, I had considerable opportunity to reflect on this and it was to my own benefit to consider the characteristics of a man who typified successful Americans. It seems to me that I would find a grouping of three characteristics: one was confidence, one was professionalism, the other was character. All those things are vaguely linked with strong personality and include reliability, honor and integrity—the kind of fellow you can depend upon when the going gets hard. And it's hard to find a better term and that's human understanding, the ability to get the best out of people and to impress upon the men you command your sincere interest in them, that they're not numbers, that they're not faceless entities on the computer but they're people. All the men that I have known have qualities which fall into those three headings one way or another. And I would quickly insist that there is no correspondence with the sub-qualities of the individual. One may be strong in his professionalism and weak in his leadership qualities of understanding people and bringing their best qualities out, and so forth.

What role does personality play?

Well, what does personality mean to you?

I mean to the point of flamboyance.

That's not personality. For example, Bradley has personality but that's of a quiet, serious man. Patton had personality in the flamboyant sense that you meant. You may find certain personalities in Bradley in Normandy and Patton in Sicily. One is the antithesis of the other, but I've always pointed out that both were great commanders. They got results but they got them in different ways and by different personalities. General Patton, I'm convinced, developed his personality deliberately as a gimmick, if you will, because he was a very serious student of war and leadership and he felt that a commander must be visible to his troops, like Henry of Navarre. So, when they would go into battle they would say, "There's the old man leading us." That explains his personality a great deal and the things that he did.

Does a soft spoken introvert have the same options as the extrovert in terms of personality?

Well, I think that you can be an introvert and still be a Patton, if you will. To me an introvert means someone who communes largely with himself and draws a great deal of his company from himself. Patton may well have been an introvert. I don't think he was but he was a quiet man when he sat with his family, not in a place where he was on "official duty." I think it's possible, probably not easy. I think a commander, to a certain extent, is a bit like a politician in that he doesn't shake hands with people but he's constantly thinking, "My men are watching me, I have to do this, I can't do that, that would suggest weakness and uncertainty and indecision or set them a bad example." All of those things come up in his mind.

Does the Army discourage individuality?

Of course not. Quite the contrary. In the case of the officer corps, if you're wise you develop some individuality in your squad, try to get the best out of every soldier in your command. When you have a small unit you can do it by the man, by knowing his name and knowing his family, his attributes, his background—sizing him up.

General, would you say over the length of your career you were more of a commander than a manager? Is it possible to make a distinction?

What do you mean a manager? It never occurred to me I was a manager.

Well, a lot of individuals claim that a staff officer is a manager whereas a field commander...

Why do you take these terms out of business? It's not a business we're talking about; we're talking about waging war. You talk about a commander or a staff officer. I know it's a modern term, one that suggests that the techniques of running a warehouse are not the techniques of running an army. I find it hard to imagine the term manager applying to anything strictly military. There are plenty of non-military activities—the business of the unfortunate.

Sir, how did you reward your people for an extra effort?

I'll just tell you, you just asked society's question. I'm a commander. When I see a soldier do something well I tell him, "Good work, Bill, good work." Just as any other, it's fairly common sense, no tricks in that. You do what comes naturally. What comes naturally is when you see a man do a hard job well you ought to say something. On the other hand you don't go patting them on the back too often cause they'll expect it. You can cheapen yourself if you're too quick to give congratulations.

General Taylor, how did you find out what was going on in the lower ranks?

Oh, don't ask this childish stuff. That's what a commander is doing all the time. That's your business. I was doing my business. I constantly was with my troops as much as you can be. It all depends on where you are. You know, being an army commander, unhappily, you can't go down and see much of the troops themselves. You should visit your next echelon and farther down with a man in the command. The division commander, for example, is with the troops constantly. It just depends on the time and where you are—your geographical distribution. Vietnam was very difficult for our commanders because the troops had never really had a division commander because there was never a division together. We had 'em scattered all around in various areas so that he never had the opportunity I had in, say, Korea. I could visit a division very easily because they were close physically and because of the close proximity to all components of it.

General, is there any such thing as a born leader?

I wouldn't know. I don't know what that means. I know we say it all the time. There's no question certainly some people have more of a quality that comes by heredity or by early associations and aptitudes which others don't have. You see it when they come in at the Military Academy. One of the purposes of the four years at West Point is to get you to give the young men exposure to environmental influences which should tend to correct deficiencies in their past and give them roughly a starting point of equal advantage.

How is West Point able to take people from such diverse backgrounds and produce individuals who, for the most part, after four years share many similar attributes of command—mannerisms, posture, that sort of thing? Is it the discipline that they have to live through?

Yes, and it's also a closed environment, deliberately made so—a monastic environment. Not so much now as in my days when I was a cadet. You never left the Military Academy. You went in in July and you never left except to go to a football game—the Army-Navy football game—until a year from the next Christmas. And there are only two ways to get into West Point; one's by the west shore railroad and one's by a little wagon trail that went over the hills into New Jersey. So it was truly a monastic life. Just as the church creates the environment around you, the importance of discipline, the importance of the armed forces, the contributions of your past—the laundry line is a very real thing for the corps of cadets. Anything around the place is designed to drive those things home. It has a great effect. I've often thought, for better or for worse, going to West Point was like being dropped on your head when you're a baby. It certainly does something to you. Some people argue that it does bad things to you, but it all depends on how you interpret the consequences.

Do West Pointers still have the edge in promotions and assignments, that sort of thing?

It ought to be apparent to the critics that no officer ever promotes anyone or accepts anyone into his official group of helpers—anyone upon whom his own success depends—unless he thinks they're the best he can get. So the idea that the West Pointer only goes out and looks for West Pointers, that's nonsense. I taught for five years out there at West Point, almost four years as superintendent, and that's a lot. I have a number of friends I know well and a tremendous number are West Pointers. So, I picked Smith not because he's a West Pointer but because I know Smith. He's a damn good man. And there may be a Jones just around the corner, ROTC, just as good, maybe even better. I just don't know him. But the reverse is just as true. Your impression of a cadet who's just a bum is a very real and living impression, too, and while we'd like to say that all the bums are thrown out before they graduate, unhappily that isn't true. So the fact is it works in reverse—it's a two-way sword. You can get no place because the people that knew you in those formative years decided you weren't any good, and you can't be any good now.

General Taylor, when you combine the element of being known by the right people with the Army's tough promotion policies, it means that a lot of really talented officers will never pin on stars. Isn't that true?

Well, yes, it's very, very tight. I don't know what it is now but it used to be roughly out of ten colonels one had a chance to become a brigadier general. And yet I look at those men's records and they were excellent. Excellent. If they got as far as colonel they've been through narrowing bottlenecks all the way to the top. And there's no question that there's a lot of luck in it. Don't ever forget the factor of luck. Very, very important. First, luck in your associates. As a young officer being around people who are going places. You get to know the people who are just above you, places like that. The kind of assignments you get; maybe a man gets a weak assignment or an easy assignment—it's no fault of his own—but if he came to me his record wouldn't look nearly as good to me as his colleague who has had the chance to prove himself in the better job. Luck goes all through this business.

General, how do you explain heroics?

What do you mean by heroics?

Well, sir, I mean for example a docile man who is placed in a combat situation and suddenly does things that he never dreamed he could do.

You could write another book on this subject. In World War II I had ample opportunity to observe and form my impressions. I was a bit disappointed when, after talking to lots of people and lots of officers and getting their ideas, there's a general agreement that hell, they'll do amazing things. They weren't thinking about making the world safe for democracy or freeing the world from Hitler or *The Stars and Stripes Forever*—they weren't thinking about much of anything but their self-respect, pride in themselves and pride in their organization. A soldier said, "I couldn't be chicken in front of a dog company." (Laughter) The opinion of a dog company meant enough to him that he did things that, left to his own devices, he never would think of doing. This doesn't explain all heroics of men but it's a good clue.

Does a commander have to take certain steps to instill loyalty in his people or does it come through the process of command?

No, loyalty doesn't come with command. Obedience goes with command. Loyalty, of course, is a relationship that can be established with a commander and men; in the lower ranks it's not particularly difficult. I've always said that the man who has the hardest job, in being the officer and the leader, is the platoon commander who lives and sleeps on the ground with his men. It's like sleeping with your valet. Its very hard to be a hero to every man in your platoon when you're constantly with them; the familiarity detracts. On the other hand, when you become the division commander you're really so far away physically from the men who are going to carry on the battle that it's very hard for that loyalty to exist. You're just up in the ozone, too far removed from the troops.

At what point, General Taylor—if at all—does aloofness play a role in an officer's approach to command?

Aloofness? Well, I don't think aloofness should ever be done as a deliberate tactic. You're aloof because you're an Army commander and you're working like hell on your daily tasks. Listen, you don't have the time. At one time I had a million men. How can I, the Army commander, have any influence or real impact on the individual? You can't. You work around the clock, doing the things you have to do in your task as Army commander. You're not aloof, you're just not available.

Isn't it something of a tradition in the Army that the airborne troops possess great esprit-de-corps and express that with more exuberance than their counterparts in other branches? Did you see this during your long tenure with those guys who actually enjoy jumping out of airplanes?

I've heard officers say, who are in the business, they never knew of a division that saluted well that did bad when they got overseas. (Laughter) I'd like to have someone verify that—it would be an interesting fact if it were true. My paratroop troops saluted all their officers with the greatest enthusiasm. They wouldn't salute anybody who didn't wear parachute boots. (Laughter) That was a club within a club—it was the airborne. They had more soldiers, usually slightly drunk, thrown into jail for not saluting officers in a neighboring unit. I'd have to go down and bail them out. (Laughter)

Looking back on your career, General Taylor, what single assignment had the greatest impact on your great success in the Army?

I find that awfully hard to answer. I suppose my assignment to Leavenworth to the Command and General Staff School from 1933 to 1935 was probably the most useful single period in terms of starting other things in motion. That was the year that General MacArthur was Chief of Staff. He knew the war stalemate was holding back a generation of officers. It took me 13 years to get to be a captain. It took Twining and Wedemeyer 17 years to get it. Meanwhile, the school system—Leavenworth—was in this Army that had no troops to command and was of tremendous importance. The school was responsible for creating the senior officers of the Army in World War II. But hardly a one had had any real command. Eisenhower had commanded a battalion of tanks in the United States in World War I, Wayne Clark had commanded a company at one time and that was typical of most of these officers. You see, this was a seniority thing—purely seniority. You had to wait until the guy died to get a billet, to be recognized in any important way. You couldn't get promoted because not enough people died.

So MacArthur in 1933 suddenly upset the Army by authorizing a few lieutenants to be in that class which up until that time had been a minimum of colonels. I was one of the six lieutenants who went in there in that class, and it brought me close to a group of people who later were to be close to the generals in World War II. That setting, there was an enormous advantage to it. Then my service with General Marshall was a great help. It got me in, assigned to the airborne business and I was out on my own with the airborne in World War II.

I don't have any more questions, sir.

Well, you're very businesslike. Are you asking the same questions of everybody?

Well, with a little variance, but not much.

I'm just curious. Did I vary materially from the kind of replies you get from these others?

Yes, sir, no answers are the same.

I'm curious. I would suspect that there's a right flank and a left flank stream in most of your interviews but there would be a hard core of great similarity in terms of answering most of those questions.

The reactions are different. For example, the question about finding out what's going on in your command. I remember one four-star general, a second or third generation West Pointer, who told me he routinely donned civilian clothes and walked outside the post to try to find someone to talk to.

You mean in disguise?

Well, incognito.

For Pete's sake.

And I have had other generals who told me they would no more think of doing that than think about jumping off a 12 story building.

Well, I'd never think of that either. (Laughter) Doing it that way, my goodness—I'm the Old Man. This is the post that I'm commanding. I don't want to talk to all these people and see about their business. I want them to know I'm the Old Man! I'm not the ice man or something like that going around the post. I don't want any gossip. I want to go about my way and establish a rapport so they'll answer my questions. I never had any trouble; GIs in my experience are the finest people in the world and they leave me breathless sometimes.

General Otto P. Weyland

A graduate of Texas A&M, "Opie" Weyland was born in Riverside, California, on January 27, 1902. After earning his wings at Kelly Field, Texas, he served a variety of assignments in the U.S. and Hawaii. Shortly after the beginning of World War II he was promoted to colonel, rising to major general at war's end. During the Korean War his competence in tactical warfare was demonstrated in ten major campaigns. While serving as commanding general of Far Eastern Air Forces, Weyland was promoted to four-star general on July 5, 1952. In May 1954 he was named commanding general of Tactical Air Command. During his five-year tenure General Weyland left his mark on numerous TAC programs and policies. He retired in July 1959 and lived in San Antonio until his death there on September 2, 1979.

23

"I WAS THE BEST IN THE BUSINESS"

A conversation with General Otto P. Weyland, USAF
February 11, 1978

You retired, golly, it's been about twenty years.

I retired from the Air Force in 1959, then I went with industry and I was with industry for about 11 to 12 years. I was with Jim McDonnell—first, McDonnell Aircraft, then McDonnell-Douglas. I was with him about 11-and-a-half years. I was with Aerojet General Corporation, oh, maybe five years. Then I was with several different software companies during periods of time as a director and consultant and so on. Then I started having some heart problems a few years back and decided that I should quit. I was doing a lot of overseas traveling. I tried to run it just the way I did when I was in the service, which was bang, bang, bang, fly all night and work all day and stuff like that. It was pretty hectic, so I decided to call it off.

When you go back to the Pentagon, for the Chief of Staff's annual meeting of all the four stars, do you find fewer and fewer people that you knew?

Well, I still know all of them in that particular gang, but the people on active duty are beginning to thin out pretty much. Bob Dixon was one of my boys. I just saw a day or two ago that he's going to retire this coming summer. And others like him. So there aren't too many left. Of course Brown, who is Chairman of the Joint Chiefs, he worked for me during the Korean War and I've known him a long time. I guess he'll probably be getting out. And Jones will be getting out, although I never did serve with him directly and vice versa. So they're beginning to thin out very, very rapidly. As a matter of fact, they're just about all gone, just a few exceptions.

Is it important to maintain these contacts, these relationships?

Oh, yes. When you quit the service you kind of rig up a new circle of friends, partly military and partly civilian, so you broaden out. I'm still fairly active socially around here and things like that. I lost my wife a little over three years ago and that made me sort of an eligible single guy who wears pants. I'm going to put in for extra-hazardous duty pay. They keep you pretty busy. As you may or may not know, every woman in the country has a particularly close widow friend and they look at you as if, boy, I'm going to fix my friend up. It's pretty wearing. But it's kind of fun.

I'll bet it is. General Weyland, looking back on your career, what would you like to be most remembered for in the Air Force?

I guess for what I did, to tell you the truth. In my career I did all sorts of things, did lots of staff duty, did lots of line instruction and things like that; but when it came to combat I always tended to line myself up when a war came along to kind of be available and there when it happened. Well, I managed to do that but I was always in the tactical business, tactical air operations. So I was pretty successful at it. I had a good basis, I knew the Army forwards and backwards. I lived with the Army at Ft. Sam Houston for two and a half years and they trusted me and I could get along with them. At the same time I protected the Air Force because I believe in the centralized control of air power and the Army was traditionally always trying to nibble away at it and get some back under their control. Well, I handled that situation pretty well with no particular trouble, so I thought it was a hell of a lot of fun in World War II and the Korean War. That was my business and I think I did it pretty well. I got a great deal of satisfaction out of it, I had a hell of a lot of fun, and I think I paid my way. Therefore, I wouldn't want to change it really. The tactical air business is the most fun there is in the Air Force.

General, what about your days at TAC? Aren't there many things that you were involved with and you initiated that are now pretty much standard practice?

Oh, yeah, sure, sure. I was pretty much the father of a lot of those tactics techniques. Of course, it's advanced, I mean technologically predominantly, but the basic principles that sort of evolved principally during World War II are much in practice. Yeah, I helped evolve those. No person invented anything, you know. With a thing as big as the military everybody gets involved. But there are certain things, lots of things in the tactical area that I pushed and were adopted and have become standard practice and accepted pretty much everywhere. Now, as I say, I claim a little credit but I sure as hell can't claim all the credit. You know, I had some awful good, damn good colonels working for me. That's sort of the secret of how you get along. Have some good guys, trust them, and if they're good and you give them a free reign and back them up, let them have one mistake or one and a half mistakes, then you kick them in the pants. Give them a pat on the back when they do well. But you have to have them in back of you. So that's what I did. I had some awfully good people and they never let me down. Well, once in a while somebody did, then I got somebody else that could do it, could hack it. Anyhow, I grew up in the racket when it was just pretty much in its infancy and I taught it over at Kelly Field and so on and so forth.

You had a reputation in the Air Force as a hard charging leader much like Curt LeMay.

I've just been reading this book, the name of it is *The Boys from Brazil*, I think it's called. Well, it's a bunch of former Nazis that have moved down to the South American whatnot. Ah, it's pretty much the

Nazi clan trying to clean out the Jews and the Jewish clans trying to stamp out the former Nazis. It's pretty interesting. It came out in this book that they mentioned a guy by name, Eric something—Rudolph, I think—who was quite a German air hero. I knew him. He was a tactical guy. He was a stupid pilot. He flew over 2,500 combat missions, predominantly against the Russians, who he cordially hated. He didn't like us very much, but he sure hated the Russians. He came over and he brought all his outfit over and landed on one of my airfields with the aircraft he had and came to see me with the offer, "Would you like to join me and we'll go and clean up on those Russians?" That was a little amusing to me at the time, although I was a little teed off at the Russians because they were holding two of my pilots who happened to land at one of their fields and I hadn't seen them—I hadn't gotten them back. So I was kind of inclined to agree with him.

When was this, General? What year was this?

This was toward the end of World War II, right toward the end. Somewhere around there. Anyhow, this cookie was all dressed up. Incidentally, he had flown hundreds of combat missions with an artificial leg. Something like this guy Bater, the Englishman. So, boy, he came in and clicked his heels and saluted smartly and had a bunch of medals hanging over him. I suppose that he had spent his life running and he could join me and we'd go and fight the Russians. So I told him, "Well, hell, don't you know—did you ever hear that they were supposed to be our allies?" But he wrote a book. A typical fighter pilot's book incidentally. You know, every fighter pilot is the best pilot that ever lived and I was one of them, so no change. But he was pretty damned good, there was no doubt about it, I'm sure of that. Now he became persona non grata in Germany after the war. God he was huffy. Well, he was just worse than huffy. But I think he was probably a pretty strong Nazi, too. You know, the Nazi Youth type of thing, great admirer of Hitler. So he didn't do very well in his native country. He went to South America as I understand. This book quoted him as being part of their organization of this Nazi group. It's supposed to be fiction but they had his name right out—all three, his first and middle name and his last name were precise and it kind of fits the guy. Well, that just

Interviews With Distinguished Generals & Admirals 245

came up, because I just finished the book last night. There were other Germans with more stature than that. Field Marshall von Rundstedt. I interviewed him for several hours. That was very interesting.

I can imagine it was. What kind of person was he?

He was obviously just a fine soldier and a fine gentleman, I'm sure of that. He was one of those—call him a Prussian, whatever you want to, but he was a complete military man and, I think, a complete gentleman. And he was quite, quite able. I had some awful good quotes from him because I asked him about how he felt about the employment of tactical air. He said, well, he never did understand how it was that I was always able to be everywhere at once, apparently. Well, I explained to him how we were organized, that the centralized control and the complete flexibility within the range limitations of our aircraft was extended by other air fields and that we could call them in the air and shift targets right as the situation came up. I gave him a little lesson in that. He sat back and said he thought it was something like that. He sure saw how it was and that he didn't have quite that good a control over the German air force. He went on to tell me it would take a little time to really express his opinion about tactical air power as we had practiced it against him. He said, well, I knew that once you started out in company ranks you could become a division commander and, as all army types are, if you get a division, boy, you think you're the king of the mountain and you have a tremendous responsibility which, of course, they do. But as a division commander, he said, of course he'd like to have about at least a squadron of fighter bombers or something of that nature. You see, that's for his immediate control, because they would help him so much and extend his artillery and so on.

Then he said he became a corps commander. Now, he said, as a corps commander he had more strategic responsibility and was getting a little smarter. As a corps commander you'd have two or three or maybe four divisions. I said, now under those circumstances if you had been a corps commander, why you'd want to have some tactical air assigned to you for personal use. But, of course, he said, just as he kept the long-range corps artillery under his immediate control and not assigned to divisions, why naturally he would keep the tactical air under his immediate control and not parcel it out to divisions. Well, he went on, now you can imagine how I felt when I was an army commander and how I'd feel as the overall commander. So my story, when I lecture at the war colleges and stuff is, the customer is always right. Well, the Air Force concept of the employment of tactical air was—he backed it up 100 percent. So it makes good reporting because he was the opposition and he was one of our customers.

If I recall correctly, he either was not a very strong Nazi or he never joined the party.

No. I doubt if he was a party member. I'm not sure about that, but, no, he was not a strong Nazi. I don't think he ever joined the party as far as I know. As a matter of fact, he was fired once or twice. He was fired at the end by Hitler, because he was trying to be honest and he would argue against Hitler and that was sort of non-habit forming.

I guess so, but he managed to survive when so many of the others didn't.

Yeah, unlike the guy up in North Africa....

Rommel?

Rommel, yeah, they polished him off. I guess he was quite a soldier. I never knew him.

While we're talking about success, what would you say contributed most to your success to getting all the way to four-star rank?

Well, there have been many factors, of course. You have to have a little luck. You have to be in the right place at the right time. You have to be available. I was available. I tried to make myself available for where I wanted to go which was into combat. Now, the back of that to start out with, early in my life—whether I chose it or not—anyhow, I became specialized in the tactical business of working with surface forces, ground forces particularly, Army especially—we were part of the Army. I worked with them and I knew them forwards and backwards. I did division and corps maneuvers with the Army here in Texas back in the Twenties, day in and day out. I watched them from the air and I'd land out in the brush and attend the critique. So I knew the racket. I knew them. I knew where they were supposed to be and what their problems were and I appreciated them and suffered for them and whatnot. So I knew the business and I got to be pretty damned good at it.

Then I went through the tactical school at Fort Leavenworth and those things—the requisite schooling, as it were—which is you don't have to, but it doesn't hurt if you do. The thing was that I had pretty well specialized and I knew my business. First of all you have to know your business and I was a very active pilot all the time. I did my own flying. When I had units under me, I flew the same types of aircraft they had. Well, the kids liked that. So when you send them in to fight, they kind of like to know that the old man knows what their problems are. The main thing is to know your business, which in my business was predominantly the tactical air operation of the Air Force, the other being the strategy of the air defense. The air defense is fighters, which we do when we're overseas in combat. You do the air defense as well as the tactical. So, that's the main thing –learn your business first. Then trying to arrange it, if it's at all possible, to be at the right place at the right time when the opportunity knocks and then when there's some combat opening up, kind of be available.

I've seen a lot of people who were probably pretty able and then all of a sudden they had family problems and they don't think they ought to go yet, they want to hang back. There are presently quite a few people like that. Well, my family knew well that I was going to go anyhow, so they'd give me a kick in the pants and say, get going and I'd cross my fingers. Okay. Then you get into the racket. Sure. I've boasted a little bit. I knew my racket and I did. There was no doubt about that. I was the best in the business. Now there are others who have come along and I helped train them. Spike Momyer was one. I kind of brought him up. He got a good start. He was an ace in World War II. He was a fighter pilot but he was smart. He started at the end of the line. I envy him being an ace because I never was, never had a chance to. When I got into combat, I was already a general and they'd always ground me and wouldn't let me go across the lines. If I flew a few missions, boy, they'd slap me. They wouldn't ground me, but they grounded me from going into combat, so I envied him. He was much younger than I, so he had that start and he was just as smart as hell. I kind of brought him up. I didn't bring him up, but I polished him off a little bit and he happened to believe a lot in my principles. So he became my chief of plans and whatnot at TAC at one time. Then I got him his first star and got him his second star, then he was on his own. He made four stars with no trouble.

But the thing is, you get some good people like that and back them up if they're good and trust them, because there's nobody who's so smart that he can do it all himself. I've known some guys who are kind of tough sons of bitches and a lot of people didn't like them. In spite of that, they seemed to do okay. I won't name any names, but on the other hand, that was not the way I operated. At least I hope that they did and I think that they did, that they liked me and they wanted to back me up. They wanted to see the old man, to make me look good. I think that's one of the keys, shall we say, to my success if I had any. I guess I did, depending upon how you measure these things. I measured by the support that I got and the wholehearted, the willing support.

Boy, I preferred not to have a yes man. Although inevitably you get them, inevitably I'm sure you get them and that makes you feel pretty good when everybody agrees with you. Well, from my standpoint that wasn't always so good. So I appreciated having some guys who would shake their heads and say, boss, it won't work. It's happened a good many times. And I'd say, well, why won't it work? I'd have one of my brilliant ideas and well, there's a fellow named Bill Kennedy here in town, he's a retired major general. He used to do that on me. He'd shake his head. He was no yes man. I'd have a brilliant idea in a staff meeting or something and he'd shake his head. I'd say, "What are you shaking your head about, Bill?" He'd say, "It won't work." Then I'd draw him out. "Why won't it work?" It kind of hurt my feelings a little bit but I wanted to know. Maybe he was wrong or maybe still, on the other hand, I still felt I was right. But, anyhow, there were times when I said, "I don't think it will work either. I think you've got a point." So, you've got to have a little honesty or you've got to have support and you've got to have honest support which I think I had. And that's the way you get by. Some guys to make you look good and back you up. There's so damned much to be done and a bunch of staff people can help you or they can hurt you. If they support you and back you up, fine. Then they'll give you good honest support. If they're scared of you, if you rule through fear, well, I don't know what will come out of it. I know some people who ruled by fear and did it rather successfully. I wouldn't like that. That isn't the way I did it. There's some people who actually operated that way and did it okay. I liked my method better.

I went through the Korean War, that was different. You see, I fought one all-out war and then I fought a limited war, a brush-fire type of war. I invented this—well, again I didn't invent it, but I almost did—what I called mobile air task forces to be ready to move anywhere in the world quickly. I practiced that. In TAC we didn't have any money, you know. I had to beg, borrow and steal and had to have my own tankers and had some old B-29s initially and converted them into tankers. Eventually, we worked on up. Anyhow, when sort of a brush-fire situation appeared in the Middle East, hell—I had about six squadrons over there the next day, because we were ready for it. I had a complete staff ready to go too. It was a racket.

General John A. Wickham

A member of the class of 1950 at West Point, John Adams Wickham, Jr. was born in Dobbs Ferry, New York, on June 25, 1928. His assignments prior to becoming a general officer included two years as aide-de-camp to General Harold K. Johnson—Army Chief of Staff—and command of a battalion in combat in Vietnam. Following nearly three years as military assistant to three secretaries of defense—Richardson, Schlesinger and Rumsfeld—General Wickham commanded the 101st Airborne Division and served as Director of the Joint Staff in the Pentagon. Promoted to four stars in July 1979, he served as Commander in Chief, United Nations Command in Korea, Vice Chief and Chief of Staff of the Army. He retired in 1987 and resides in Tucson.

24

"I WAS ALMOST KILLED IN VIETNAM"

A conversation with General John A. Wickham, Jr., USA
September 5, 1990

Are we running out of combat-tested leaders in the Army and is there enough of a reservoir of leadership potential to provide capable leadership in the future?

The answer to the reservoir is that we are running out of combat leaders. In Grenada only 11 percent of the Army troops that went in there had ever heard a shot fired in anger. That does not, I believe, say bad things about the potential for leadership. If you take a look at the interwar period, between World War I and World War II, the Pattons and the Eisenhowers, the Ridgways, were in the Army. They were the ones who had troops training with broom sticks because they didn't have enough weapons. Yet out of that interwar period, the officers who stayed with the system, came the renowned leaders: the Pattons and the Bradleys, the Marshalls, and some less renowned but equally capable, J. Lawton Collins, Van Fleet, Lemnitzer, Charlie Bolte, people like that who had not had any combat experience, yet proved to be very capable leaders in the war.

The extent of training that is under way today, the focus on mentoring of leaders, the simulation activity, the war gaming activity, the emphasis on the high level of readiness in units, all are extremely challenging in leader development, far more than at any time, I believe, in the nation's history. The national training center at Fort Irwin, where we run battalions and brigades up against the Soviet motorized rifle regiment, fully instrumented with air-to-ground, ground-to-air, ground-to-ground training activity is as close as one can come to combat without being shot at with live ammunition. It has been a very valuable training ground for our forces going, light and heavy, going into the Middle East. We had nothing like that in years past. I have seen in my 37 years of active duty a profound emphasis on mentoring and selection of leaders based on merit and based on capability. When I was a battalion commander and even a brigade commander, it was almost left to chance whether you got brigade command. But in recent years now we have boards that select people for command based on merit. We're putting the best people in. When I went to command at battalion and brigade level there was no preparation. You just slam, bang, then you're into a unit and if you don't perform you're out. We had a zero defects Army in those years, 15 to 20 years ago, which meant you couldn't make a mistake, therefore, don't risk anything, don't extend yourself, because you're going to read about it in the OER (officer efficiency report) and you'll likely get relieved and ruin your career. That's gone today.

General, what sorts of things did you do to bolster leadership in the Army during your years as Chief?

When I was Chief, every month I went to Fort Leavenworth and the months I couldn't because of a JCS meeting or a trip, I'd do it by video-audio link up. I would talk to all of the new battalion, brigade and division commanders, about 80 of them a month. I'd spend a couple of hours with them giving them my philosophy of command and issues in the Army and then questions and answers back and forth. I had none of that when I came up through the Army. Having seen the change that's taken place, having seen the quality of these leaders firsthand in the field, I cannot help but be convinced that the Pattons and the Eisenhowers and the Ridgways are in our Armed Forces in great numbers and a war will give them the publicity that they don't have today. I'd hope that we don't have the war to give them the publicity, but they are there.

I'll give you a specific illustration: rangers. We've created a ranger regiment in the Army. The ranger battalions were first created by General Abrams when he was Army Chief of Staff. In order to become a ranger battalion commander, you had to have commanded successfully, very successfully, a battalion of armored forces or of infantry forces. In order to be the regimental commander of the ranger regiment you had to have successfully completed command of a brigade. That's the only regiment we do that for. Those are the only battalions we do that for. It's a very rich command experience. We couldn't do it throughout the Army because you just run out of time and you run out of command opportunities. But we do that because we're trying to grow leaders in a very critical area of highly deployable, lethal, special operations forces.

Can you walk me through the mentorship process as it occurred in your career, General?

Well, I had very little mentoring that I look back on. I would have liked to have had more. I would have liked to have had senior officers call me in and say here are the good things that you're doing, Wickham, here are the bad things you're doing, and here's how you can grow. What has transpired in the Army I guess in the past 10 years or so has been an evolution of emphasis on the leader to lead. That the leader has a moral obligation to mentor and counsel those who are coming along. I've literally talked with tens of thousands of soldiers and officers in my years as Chief and I did a lot of video tapes that were shown to virtually everybody in the Army, active and reserves. Many of those tapes focused on the issue of mentoring and counseling young leaders. That this was an obligation to transfer your experience and that the kind of positive leadership we're looking for is not the negative dictatorial leadership that we sometimes have associated with the past. The hard-drinking, foul-mouthed, profane leaders that were harsh in dealing with subordinates. Like the zero defects mentality. We tried to purge that from the Army and to foster positive, inspirational example-setting leadership. That's hand-in-glove with the mentoring.

Another illustration, in my responsibilities as Army Chief of

Staff—I think all the service chiefs had the same responsibility—they are responsible among their other duties for managing the general officer corps. You basically have a battalion of generals, 412, and the Army Chief makes all of the assignments. He develops the stable of upward-mobile opportunity for the more capable people. I guess I spent on the average 20 to 25 percent of my time just managing the general officer structure. And I needed all the help I could get in doing that and therefore went out to solicit views, peer ratings, peer views of brigadier generals and major generals to determine how people were viewed from below. It was very, very revealing. It was another factor in making decisions about assignments, because I would find out whether some individuals were ruthless with subordinates. In that case, I would obviously be reluctant to give an individual like that a position of responsibility over soldiers and over younger leaders.

In my guidance to promotion boards—the promotion boards to select brigadier generals and promotion boards to select two stars from the one stars—one of the points I made in all of my oral guidance was we need to be concerned about selecting leaders who set the positive example of leadership and transfer knowledge and that do well at mentoring. So I give all of these little snippets to give you a sense of how the culture has changed over the years in the way of developing solid leadership qualities and thorough leadership training throughout the Army. Because we all recognize, absent of wars, we're going to lose combat experience. I'd had no combat experience before I went in to command an infantry battalion in Vietnam. I was almost killed in Vietnam. But I think I led very effectively. I was awarded two Silver Stars. I was recommended for the Distinguished Service Cross for combat action. I went into an air mobile unit, the 1st Cavalry Division, that was rather unique and challenging in the way it conducted tactics, its doctrines, the use of helicopters. You had to think very fast, you had to think in terms of 90 knots as opposed to two and one-half miles an hour to move forces around on the battlefield. Some people were able to do it, some couldn't. There's no training you could get, I don't think, that would prepare you for that

General, at the end, the very end of the movie *Patton* the two characters are talking just before they have dinner. Bradley turns to

Patton and says, "I have a feeling from now on just being a good soldier won't mean a thing. I'm afraid we're going to have to be diplomats, administrators, you name it." How prophetic was that?

I'm sure that Eisenhower and Bradley and Patton, were we to ask them today, would have said, oh, yes, if we're going to have peace, it has to be peace of the strong. In order to be strong, you have to have solid soldiers, well-trained just the way we tried to be well trained before World War II. I don't think there's anything inconsistent with that. I don't think they are saying that what one needs to do when you get out of West Point or Annapolis or the Air Force Academy is to put aside soldierly or military things and become a politically-oriented individual. Colin Powell is an illustration. Colin is thoroughly grounded on military matters. I've known Colin many years. He was one of my brigade commanders. He is very capable on the military side but he's also been thrust into politico-military affairs and has performed superbly there. But he hasn't set aside the military issues and I think the blend of them are working very well in the senior positions that he's held so far and is likely to hold in the future. But Colin is unique. I've got a young son-in-law who is a major of infantry. He is better as a major of infantry—as I look back on my own life—he's more capable as a soldier, as a potential leader at battalion and beyond than I was as a major. It's because of the rich training diet that is available to them at that level and the thoughtful attention by senior leaders for many, many years—the four stars, the three stars, all of the general officers—to build a solid reservoir of trained leaders.

As I said, I was almost killed in Vietnam. I don't carry it on my sleeves in any way but I feel I was given another chance in life. I know I was given another chance in life. I was given the last rights in Vietnam. For that reason I felt I must do a little bit more, try to leave more of a legacy than I might otherwise had done. I think every senior leader has felt similarly, that he wants to do right by the position of responsibility that's been entrusted to him. Leave it better than he found it, to try to leave a legacy of inspiration in the people that he's touched. People therefore who would be critical of themselves or others are missing the fact that there is a motivation to try to leave the place better and to try to leave an inspirational example for the future. I think I tried to approach that during my four years as Army Chief.

As the Armed Forces continue to downsize because of budgetary concerns, obviously the number of people in the Army is going to shrink. At the service academies they're going to continue to put out well-trained, competent officers. Increasing numbers of those are going to be women who don't have a combat role in the Army. Does it make sense to you, General, that we have women in an Army where they can't fight?

Women will be in combat. They have been historically and they have been recently and they will be in the future—women in infantry or armor roles where they direct fire power against the enemy; that we don't have because of congressional constraints. The Army's followed the pattern of the Air Force in that regard. You don't have women as fighter pilots. Perhaps one day we will. But that does not mean that women are not in combat and will not be in self-defense taking up weapons against the enemy and killing the enemy as they would be in reserve areas or as they would be if they were in a signal unit or an ordinance unit down in a brigade and were ambushed. They'd have to defend themselves and they will do well at that. The reality is that we need women in the armed forces. That is a talent that we probably have not capitalized on adequately in the past. The 10 percent of the service academies that are female is a reasonable level. They don't stay in the service at the same rate as men do because I think the women basically have found the services, some of them, to be a less tolerant place than they had hoped for. But that may change with time. So I'm a supporter of women in the armed forces.

General Wickham, in the midst of downsizing, the civilian side of the defense slate seems to have been ignored, if not spared. What do you feel should be done, in terms of changes, to make the civilian-military team work better and perhaps less expensively?

When administrations change they will bring in their own coterie of civilians to work with them and I think there's a healthy phenomenon that takes place there. New insights, challenges to the way we've done business and the ability to carry some of the freight with the congress. It may be easier for some of our civilians to go over and argue with the Congress than sending the military up to the hill all the time. So I think they provide a healthy interface with the military. But it is cumbersome and the military just has to grow comfortable with that. How would we change it? I would say one way would be for the President on down to encourage staffs to be as lean as possible. There is a tendency for staffs to grow. I think Mr. Cheney has been quick, as we started to cut back on the Defense Department, to try to reduce his own staff unilaterally and to reduce the number of flag officers there. That is a healthy phenomenon. I'd like to see the same sort of thing out of the Congress. The staffs in the Congress have grown like topsy-turvy. Yet they insist that others take the cuts first. So leanness of staff is one important direction.

I think the second direction would be for the civilian leaders—and I've seen it as I said in OSD—to develop much more faith and tolerance in the military staff capabilities and to rely more on them rather than to try to duplicate staffs. One of the downsides of civilian structure within our system of the Department of Defense can be suspicion: we-they. You people in the military are part of a guild and we are not. And we're not sure we entirely trust you. I'm going to have my own people looking out for events here. Somehow they've got to get over that lack of trust and suspicion and develop a better awareness. My guess is it's going to be difficult because we have fewer and fewer civilians now from which we draw our leadership—it's true in Congress—who have any military experience at all. Therefore, the suspicion is likely to exist in the future. Somehow or other, if we can build a better tolerance of each other—that's a cultural thing—but that would be helpful. I think some of the initiatives that have grown out of the Goldwater-Nichols Act probably are very healthy. There's more professionalism in specialties. The idea of more professional capabilities in the acquisition business, the procurement business, makes a great deal of sense. I think efficiency is going to grow out of that. Procurement today is far more complicated than it was in years past. It's going to grow increasingly more complicated in the future. Professionals there will be beneficial.

I think a fourth area is that we must somehow or other figure out how we can capitalize on experience. The ethics acts that have grown up in recent years, which disable people from dual compensation, which disable people from coming back to work in the government or going to work in the defense industry, I think, tends to reduce the opportunity to capitalize on experience. You've got to find the right balance clearly between the revolving door, where people make use of their contractual knowledge for their own personal gain, and a balance to minimize that and yet at the same time capitalize on experience. I don't think we do an adequate job right now, because of the constraints of the ethics act, it is very difficult to get high quality people to come in at middle levels at the civilian side of the government because of the baggage they carry when they leave the government about adequate employment in the civil sector. You're either going to get people who are retired from the business world, very elderly gentlemen like Jim Ambrose who was worth his weight in gold—unfortunately there's not many Jim Ambroses around, he was Undersecretary of the Army, for example—or you're going to get people very young that may not have the kind of experience that the Department of Defense needs. Somehow or other we've got to fix the ethics act prohibitions that enable us to attract more and more high quality, higher quality in the middle-level civilian leadership.

General Louis L. Wilson

A member of the class of 1943 at West Point, General Wilson was born in Huntington, West Virginia on January 10, 1919. During World War II he was assigned to the 358th Fighter Group in the Eighth and Ninth Air Forces and flew 114 combat missions in a P-47. Following the war he served again with Eighth Air Force and in 1964 began a seven-year association with ICBMs. He was the Air Force Inspector General prior to appointment as Commander in Chief Pacific Air Forces in 1974. Thereafter he resided in Tucson, Arizona, until his death on June 25, 2010.

25

ADVICE TO COMMANDERS— TEN POINTS

A Conversation with General Louis L. Wilson, Jr., USAF October 22, 1986

General Wilson, you have a reputation within the Air Force for your common-sense approach to leadership. Would you please give me a background into your famous "Advice to Commanders"?

For each new crop of selectees to brigadier general they have a two-day session in the Pentagon to brief them on their new status and more or less tell them what is expected of them. Charm school is what it's called unofficially. At the time I was the Inspector General of the Air Force, and along with all other key Air Staff members, we were called on to talk to the new brigadier generals to explain our role in the scheme of things. They had there in my office a canned speech given for years by my predecessors which dwelt on what the IG did and so on.

The evening before I was to speak, while on the way home from an inspection on the West Coast, I first looked at the canned speech and decided it wasn't at all what I wanted to say to a new bunch of brigadier generals. So I jotted down some thoughts I had gleaned over the years about success and failure as a leader. There were nine of them at the time. The next day, after I had made my talk, my office started getting calls for copies of what I had said. It was a bestseller and I still get calls for copies.

A year or so later, when I went to PACAF, I saw a need for guidance for commanders all up and down the line. So I revised my thoughts on leadership, added the 10th point on integrity and distributed to all concerned. It seemed to help.

1. BE TOUGH. Set your standards high and insist that your people measure up. Have the courage to correct and, if necessary, chastise those who fail to do so. Discipline those who won't conform. In the long run your people will be happier. Almost certainly morale will be higher, your outfit better, and your people prouder. Good outfits have tough commanders—not arbitrary or unfair or cruel—just tough.

2. GET OUT FROM BEHIND YOUR DESK and see for yourself what's going on. Your place of business is where the action is. Leave your footprints all over the place. Your subordinates will see that you're interested in their problems, working conditions and welfare. Many of your people problems will go away if you practice this point.

3. SEARCH OUT THE PROBLEMS - they are there. If you think there are no problems in your organization, you are ignorant. Again, they are there. The trick is to find them. Foster an environment that encourages people to bring problems to you. If you shun problems you are not fit to command.

4. FIND THE CRITICAL PATH TO SUCCESS - then get personally involved on a priority basis. Let your influence be felt on the make/break issues in your organization. Avoid the "activity trap"—don't spend your valuable time on inconsequential or trivial matters. Weigh in where it counts. Be the master of your fate—don't leave it to chance.

5. BE SENSITIVE. Listen to your people. Communicate. Be perceptive. Recognize that communications are shared perceptions. Empathize. Learn to recognize problems. Seek ideas. Be innovative. Listen, listen, listen!

6. DON'T TAKE THINGS FOR GRANTED. Don't assume things wrong have been fixed—look for yourself. Neither assume they will stay fixed. The probability is high that "fixed" problems will recur. Recheck the fix.

7. DON'T ALIBI—just fix it. Remember you and your outfit can never be perfect. People will make mistakes. Don't be defensive about things that are wrong. Nothing is more disgusting than the individual who can do no wrong and has an alibi for any and everything that goes awry.

8. DON'T PROCRASTINATE. Don't put off those hard decisions because you're not willing to make them today. It won't be easier tomorrow. This doesn't mean to make precipitous or unreasoned decisions just to be prompt. But once you have arrived at what you believe is correct, get on with it. Don't stymie progress.

9. DON'T TOLERATE INCOMPETENCE. Once a person has demonstrated that he is too lazy or too disinterested, or unable because of aptitude to get the job done, you must have the courage to terminate his assignment. You cannot afford to do less. On the other hand, when your people are doing good work, recognize it and encourage them. Certainly they will do even better.

10. BE HONEST. Don't quibble. Tell it like it is. Insist that your people do likewise. They set their patterns

based on your example. Absolutely nothing can be more disastrous than garbled information, half-truths and falsifications. Make sure your people know how you stand on this matter. Encourage them to come to you if they have doubts or are troubled about veracity in the outfit. You must create an atmosphere of trust and confidence. And be honest with yourself—don't gimmick reports and figures or use cunning ways just to make things look good. If you do, you are a loser before you start.

TO SUM UP: Your task is to lead. This requires hard work, enthusiasm for the job, and sensitivity to what's going on about you. You must set your standards high, be involved, listen, know what the problems are, remove the weak, promote the strong—and to do this well you've got to be tough. Finally, remember that honesty and integrity are basic to it all. Don't risk success—practice these 10 points. If you do, you certainly won't be a failure.

About the Author

The son of an Army Air Force fighter pilot, R. Manning Ancell began his career at KOB TV in Albuquerque, New Mexico where he produced a number of half-hour documentaries on subjects as diverse as the New Mexico State Prison, the space shuttle, and astronaut Edgar D. Mitchell. One of his documentaries, "The Thief on the Street Where You Live," earned first place awards from United Press International and the New Mexico Broadcasters Association.

In 1971 he was commissioned as a public affairs officer in the Naval Reserve. His annual two-week tours of active duty included service at U.S. Atlantic Command in Norfolk, where he wrote a major speech for Admiral Ralph W. Cousins, the Commander in Chief; in the office of the Chief of Information in support of Secretary of the Navy John Warner's program for the Navy's 200th anniversary; aboard the battleship New Jersey during her re-commissioning; and with the Bureau of Medicine and Surgery in the opening weeks of Operation Desert Storm.

Ancell became publisher and executive editor of *Colorado Business Magazine* in Denver in 1979 and was later publication director of *Denver Business* magazine, responsible for its transition from a tabloid newspaper to a four-color magazine.

In 1984 Ancell was recalled to active duty as a lieutenant commander in the Navy Reserve and ordered to New Orleans, where he headed recruiting advertising and marketing. The following year he was released from active duty and became publisher of *Endless Vacation* and oversaw its redesign and launch as an international travel magazine.

First published in *New Mexico Magazine* at the age of eighteen, Ancell is the author or co-author of more than 200 articles in magazines, newspapers, and journals. He is the co-author, with General Edward C. Meyer, of *Who Will Lead?* (Praeger, 1995), and author of *The Biographical Dictionary of World War Two Generals and Flag Officers* (Greenwood Press, 1996).

In 1998 Ancell founded The Society for Four-Star Leadership, which promotes the contributions to American history of America's highest-ranking generals and admirals. He resides with his wife, Dr. Christine Miller, in Norfolk, Virginia.